SpringerBriefs in Physics

SpringerBriefs in Physics are a series of slim high-quality publications encompassing the entire spectrum of physics. Manuscripts for SpringerBriefs in Physics will be evaluated by Springer and by members of the Editorial Board. Proposals and other communication should be sent to your Publishing Editors at Springer.

Featuring compact volumes of 50 to 125 pages (approximately 20,000–45,000 words), Briefs are shorter than a conventional book but longer than a journal article. Thus, Briefs serve as timely, concise tools for students, researchers, and professionals.

Typical texts for publication might include:

- A snapshot review of the current state of a hot or emerging field
- A concise introduction to core concepts that students must understand in order to make independent contributions
- An extended research report giving more details and discussion than is possible in a conventional journal article
- A manual describing underlying principles and best practices for an experimental technique
- An essay exploring new ideas within physics, related philosophical issues, or broader topics such as science and society

Briefs allow authors to present their ideas and readers to absorb them with minimal time investment. Briefs will be published as part of Springer's eBook collection, with millions of users worldwide. In addition, they will be available, just like other books, for individual print and electronic purchase. Briefs are characterized by fast, global electronic dissemination, straightforward publishing agreements, easy-to-use manuscript preparation and formatting guidelines, and expedited production schedules. We aim for publication 8–12 weeks after acceptance.

More information about this series at http://www.springer.com/series/8902

James M. Cline

Advanced Concepts
in Quantum Field Theory

With Exercises

Springer

James M. Cline
Department of Physics
McGill University
Montreal, QC, Canada

ISSN 2191-5423   ISSN 2191-5431 (electronic)
SpringerBriefs in Physics
ISBN 978-3-030-56167-3   ISBN 978-3-030-56168-0 (eBook)
https://doi.org/10.1007/978-3-030-56168-0

This Springer imprint is published by the registered company Springer Nature Switzerland AG
The registered company address is: Gewerbestrasse 11, 6330 Cham, Switzerland

Foreword

Quantum Field Theory (QFT) is the language in which Nature's Laws at their most fundamental level are written, and this makes it as important a subject for a scientist to learn as is the learning of English for a student of Shakespeare or French to a student of Voltaire. This is why many good QFT texts have been written to help novices find their way around what is a vast and rich subject.

But real fluency is required to appreciate masterpieces written in any language, and it is equally true that a full appreciation of Nature's Laws requires going beyond an introductory treatment of QFT methods. Indeed, there is so much accumulated wisdom about QFT that learning it is a lifelong task that never really ends.

That is where this book comes in. This book is based on graduate lectures given at McGill University by the author, who is a world expert in the subject. It goes beyond the usual introductory topics found in many textbooks and dives deeper into important grottos of knowledge—such as the nitty-gritty of renormalization and a discussion of non-perturbative issues like tunnelling and theta vacua in nonabelian gauge theories. These topics have both practical and conceptual value and are often not discussed in introductory textbooks.

Most importantly, the book is chock-filled with useful exercises. Simply reading a book about a subject as vast as this doesn't really move anyone's knowledge of it beyond a level good only for cocktail-party conversation. Developing a practitioner's understanding requires using QFT tools to calculate something, but finding a calculation that is both instructive and not insuperably difficult is hard, even for old hands. A book as loaded with insight as this one is a gift to the reader. But when it is also crammed full of well-chosen exercises, it becomes a gold mine, particularly when the exercises have been tested and debugged by generations of students in the forge of their own learning.

This book is an invitation to an intellectual feast. It comes with many courses and is prepared by a five-star chef. Bon appetit!

Hamilton, ON, USA   Clifford P. Burgess
June 2020

Preface

These notes represent the second half of a year-long quantum field theory course that was given at McGill University. It assumes the reader understands the basics of field theory at the tree level. I start with the loop expansion in scalar field theory to illustrate the procedure of renormalization, and then extend this to QED and other gauge theories. My goal is to introduce the most important concepts and developments in QFT, without necessarily treating them all in depth, but allowing you to learn the basic ideas. The topics to be covered include:

- Perturbation theory: the loop expansion; regularization; dimensional regularization; Wick rotation; momentum cutoff; $\lambda\phi^4$ theory; renormalization; renormalization group equation; Wilsonian viewpoint; the epsilon expansion; relevant, irrelevant and marginal operators; Callan-Symanzik equation; running couplings; beta function; anomalous dimensions; IR and UV fixed points; asymptotic freedom; triviality; Landau pole
- The effective action: generating functional; connected diagrams; one-particle-irreducible diagrams; Legendre transform
- Gauge theories: QED; QCD; anomalies; gauge invariance and unitarity; gauge fixing; Faddeev-Popov procedure; ghosts; unitary gauge; covariant gauges; Ward Identities; BRS transformation; vacuum structure of QCD; instantons; tunnelling; theta vacuua; superselection sectors; the strong CP problem.

I would like to thank Balasubramanian Ananthanarayan (IISc, Bangalore) for his encouragement to publish these notes.

Montreal, Canada   James M. Cline

The original version of the book was revised: Missed out corrections have been updated. The correction to the book is available at https://doi.org/10.1007/978-3-030-56168-0_16

Contents

Chapter 1
Introduction

Physics, like all sciences, is based upon experimental observations. It's therefore a good thing to remind ourselves: what are the major experimental observables relevant for particle physics? These are the masses, lifetimes, and scattering cross sections of particles. Poles of propagators, and scattering and decay amplitudes are the quantities which are related to these observables:

$$\text{pole of } \frac{1}{p^2 - m^2} \leftrightarrow (\text{mass})^2, \tag{1.1}$$

$$\mathcal{T}_{a \to bc} \leftrightarrow \text{decay rate}, \tag{1.2}$$

$$\mathcal{T}_{ab \to cd} \leftrightarrow \text{scattering cross section}. \tag{1.3}$$

These observables, which are components of the *S-matrix* (scattering matrix) are the main goals of computation in quantum field theory. To be precise, $\mathcal{T}$ is the transition matrix, which is related to the S-matrix by

$$S_{fi} = \mathbf{1}_{fi} + (2\pi)^4 i \delta^{(4)}(p_f - p_i) \mathcal{T}_{fi}. \tag{1.4}$$

In a previous course you learned about the connection between Green's functions and amplitudes. The recipe, known as the LSZ reduction procedure (after Lehmann, Szymanzik and Zimmerman) [1, 2], is the following. For a physical process involving n incoming and m outgoing particles, compute the corresponding Green's function. Let's consider a scalar field theory for simplicity:

$$G_{(n+m)}(x_1, \ldots, x_n, y_1, \ldots, y_m) = \langle 0_{\text{out}} | T^* [\phi(x_1) \cdots \phi(x_n) \phi(y_1) \ldots \phi(y_m)] | 0_{\text{in}} \rangle. \tag{1.5}$$

The blob represents the possibly complicated physics occurring in the scattering region, while the lines represent the free propagation of the particles as they are traveling to or from the scattering region. It is useful to go to Fourier space:

The original version of this chapter was revised: The errors in this chapter have been corrected. The correction to this chapter can be found at https://doi.org/10.1007/978-3-030-56168-0_16

J. M. Cline, *Advanced Concepts in Quantum Field Theory*, SpringerBriefs in Physics, https://doi.org/10.1007/978-3-030-56168-0_1

Fig. 1.1 $n \to m$ scattering process

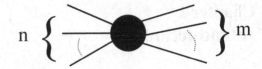

$$\tilde{G}_{(n+m)}(p_1, \ldots, p_n, q_1, \ldots, q_m) = i^{n+m} \frac{(2\pi)^4 \delta^{(4)}(\sum p_i - \sum q_i) \Gamma_{n+m}(p_1, \ldots, p_n, q_1, \ldots, q_m)}{(p_1^2 - m^2) \ldots (p_n^2 - m^2)(q_1^2 - m^2) \ldots (q_m^2 - m^2)}.$$

$$(1.6)$$

The delta function arises because we have translational invariance in space and time, so momentum and energy are conserved by the process. This equation makes the picture explicit. We see in the denominator the product of all the propagators for the free propagation. In the numerator we have the function Γ_{n+m}, called the *proper vertex function*, which represents the blob in the picture. This function contains all the interesting physics, since we are interested in the interactions between the particles and not the free propagation (Fig. 1.1).

Now we can state the LSZ procedure: to convert the Green's function to a transition amplitude, truncate (omit) all the external line propagators. In other words, the proper vertex Γ_{n+m} is the T-matrix element we are interested in. I will come back to the nontrivial proof of this statement later. First, we would like to use it to get some concrete results. The problem is that, in general, there is no analytic way to compute Γ_{n+m} if it is nontrivial. In free field theory, the only nonvanishing vertex function is $\Gamma_2 = i(p^2 - m^2)$, the inverse propagator. Once we introduce interactions, we get all the vertex functions, but they can't be computed exactly. We have to resort to some kind of approximation. Since we know how to compute for free fields, the most straightforward approximation is that in where the interactions are weak and can be treated perturbatively. In nature this is a good approximation for QED and the weak interactions, and also for the strong interaction at sufficiently high energies.

However, these realistic theories are a bit complicated to start with. It is easier to learn the basics using a toy model field theory. The simplest theory with interactions which has a stable vacuum is $\lambda\phi^4$. We simply add this term to the Klein-Gordon Lagrangian for a real scalar field[1]:

$$\mathcal{L} = \frac{1}{2} \left(\partial_\mu \phi \partial^\mu \phi - (m^2 - i\varepsilon)\phi^2 \right) - \frac{\lambda}{4!} \phi^4. \tag{1.7}$$

The factor of 1/4! is merely for convenience, as will become apparent. There are several things to notice. (1) The $i\varepsilon$ is to remind us of how to define the pole in the propagator so as to get physical (Feynman) boundary conditions. We always take $i\varepsilon \to 0$ finally, so that (2) the Lagrangian is real-valued. The latter is necessary in order for $e^{iS/\hbar}$ to be a pure phase. Violation of this condition will lead to loss of unitarity, i.e., probability will not be conserved. (3) The $\lambda\phi^4$ interaction comes with a $-$ sign: the Lagrangian is kinetic minus potential energy. The $-$ sign is necessary so

---

[1] I use the metric convention $p_\mu p^\mu = E^2 - \vec{p}^2$.

that the potential energy is bounded from below. This is the reason we consider $\lambda\phi^4$ rather than $\mu\phi^3$ as the simplest realistic scalar field potential. Although $\mu\phi^3$ would be simpler, simpler, it does not have a stable minimum—the field would like to run off to $-\infty$.

The above Lagrangian does not describe any real particles known in nature, but it is similar to that of the Higgs field which we shall study later on when we get to the standard model. The coupling constant λ is a dimensionless number since ϕ has dimensions of mass. We will be able to treat the interaction as a perturbation if λ is sufficiently small; to determine how small, we should compute the first few terms in the perturbation series and see when the corrections start to become as important as the leading term.

The tool which I find most convenient for developing perturbation theory is the Feynman path integral for the Green's functions. Let's consider the generating function for Green's functions:

$$Z[J] = \int \mathcal{D}\phi \, e^{iS[J]/\hbar}, \tag{1.8}$$

where

$$S[J] = \int d^4x \left(\mathcal{L} + J(x)\phi(x)\right). \tag{1.9}$$

Recall the *raison d'être* of $Z[j]$: the Green's functions can be derived from it by taking functional derivatives. For example, the four-point function is

$$G_4(w, x, y, z) = \frac{1}{i^4 Z} \frac{\delta^4 Z[J]}{\delta J(x_1)\delta J(x_2 \delta J(y_1)\delta J(y_2)}\bigg|_{J=0} \tag{1.10}$$

$$= \frac{1}{Z[0]} \int \mathcal{D}\phi \, e^{iS[0]/\hbar}\phi(w)\phi(x)\phi(y)\phi(z). \tag{1.11}$$

If we had no interactions, this Green's function could be computed using Wick's theorem to make contractions of all possible pairs of ϕ's as shown in Fig. 1.2. This is not very interesting: it just describes the free propagation of two independent particles. The corresponding Feynman diagram is called *disconnected* since the lines remain separate. The disconnected process is not very interesting experimentally. It corresponds to two particles in a collision missing each other and going down the beam pipe without any deflection. These events are not observed (since the detector is not placed in the path of the beam).

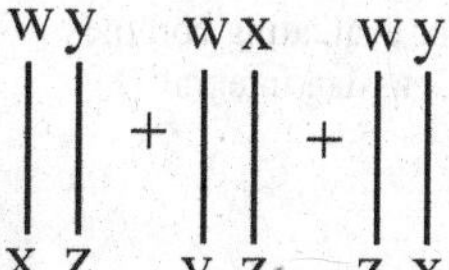

Fig. 1.2 4-point function in the absence of interactions

But when we include the interaction we get scattering between the particles. This can be seen by expanding the exponential to first order in λ:

$$G_4(w, x, y, z) \cong \frac{1}{Z[0]} \int \mathcal{D}\phi \, e^{iS[0]/\hbar} \phi(w)\phi(x)\phi(y)\phi(z) \int d^4x' \left(-\frac{i}{\hbar} \frac{\lambda}{4!} \phi^4 \right).$$

(1.12)

Before evaluating this path integral, I would like to digress for a moment to discuss its much simpler analog, the ordinary integral

$$Z = \int_{-\infty}^{\infty} \frac{dx}{\sqrt{2\pi}} e^{\frac{i}{2}ax^2}.$$

(1.13)

This is related to the well-known trigonometric ones, the Fresnel integrals, and they can be computed using contour integration [3] along the contour shown in Fig.1.3.

$$Z = 2 \int_0^{\infty} \frac{dx}{\sqrt{2\pi}} e^{\frac{i}{2}ax^2} = -2e^{i\pi/4} \int_{\infty}^{0} \frac{dy}{\sqrt{2\pi}} e^{-\frac{1}{2}ay^2} = \frac{e^{i\pi/4}}{\sqrt{a}}.$$

(1.14)

The second equality is obtained by using the fact that the integral around the full contour vanishes, as well as that along the circular arc (at ∞). Therefore we have shown that the oscillatory Gaussian integral is related to the real one. The integral Z is analogous to the field theory generating functional $Z[0]$. And the analogy to the 2-point function (the propagator) is

$$\frac{1}{Z} \int_{-\infty}^{\infty} \frac{dx}{\sqrt{2\pi}} e^{\frac{i}{2}ax^2} x^2 = \frac{1}{Z} \frac{2}{i} \frac{dZ}{da} = \frac{i}{a}.$$

(1.15)

If we carry out the analogous procedure in field theory, we obtain the momentum-space propagator $\tilde{G}_2 = i/(p^2 - m^2)$. This little exercise shows you where the factor of i is coming from. It also belies the statement you will sometimes hear, that the Feynman path integral only rigorously exists in Euclidean space.

Now let's return to the 4-point function. When we do the contractions, we have the possibility of contracting the external fields with the fields from the interaction vertex. This new *connected* contribution (Fig. 1.4) comes *in addition* to the disconnected

Fig. 1.3 Complex contour
for evaluating complex
Gaussian integral

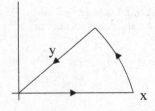

Fig. 1.4 Connected 4-point function at linear order in λ: a tree diagram

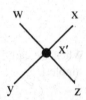

Fig. 1.5 Contribution to disconnected 4-point function at linear order in λ

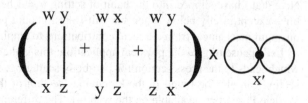

Fig. 1.6 Another contribution to the disconnected 4-point function at linear order in λ

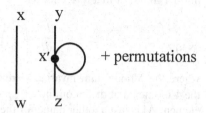

+ permutations

one shown in Fig. 1.3. In the latter, we contract the ϕ's within the interaction only with themselves. This results in a *loop* diagram, in fact a two-loop diagram. There is also another kind of disconnected diagram where one of the particles is freely propagating, while the other feels the effect of the interaction, Fig. 1.6. We are going to focus on the connected contribution, Fig. 1.4, for right now. This kind of diagram is called a *tree diagram* because of its stick-like construction, to distinguish it from loop diagrams, such as the figure-eight appearing in Fig. 1.5.

In evaluating the connected diagram, we have to take into account all the possible ways of Wick-contracting. Start with $\phi(w)$: it has 4 possibilities among the 4 fields of $\phi(x')^4$; let us choose one of them. Then $\phi(x)$ has 3 remaining possibilities. $\phi(y)$ has 2 and $\phi(z)$ is left with 1. Thus there are $4 \cdot 3 \cdot 2 \cdot 1 = 4!$ contractions which all give the same result. This is called the *statistical factor* of the Feynman diagram, and it is the reason for including $1/4!$ in the normalization of the coupling: the $4!$ factors cancel out in the physical amplitude, making the final result look nicer.

$$G_4(w, x, y, z) = -i\lambda \int d^4x' G_2(w, x')G_2(x, x')G_2(y, x')G_2(z, x'), \qquad (1.16)$$

where $G^{(2)}(x, x')$ is the Klein-Gordon propagator,

$$G_2(x, x') = \int \frac{d^4 p}{(2\pi)^4} \frac{e^{-ip(x-x')}}{(p^2 - m^2 + i\varepsilon)}.\qquad(1.17)$$

When we carry out the LSZ procedure to find the proper 4-point vertex, it is very simple:

$$\Gamma_4 = -i\lambda.\qquad(1.18)$$

Note that I have slipped into the habit of setting $\hbar = 1$ here. We will work in these units except briefly below when we will want to count powers of $\hbar$ to distinguish quantum mechanical from classical contributions to amplitudes.

Let's pause to give the physical application: this is the transition matrix element which leads to the cross section for two-body scattering. The formula for the cross section can also be found in the kinematics section of the very useful Review of Particle Properties, available on the web [4]. The differential cross-section is given in Eq. (48.33) of that reference as

$$\frac{d\sigma}{dt} = \frac{1}{64\pi s} \frac{1}{|\mathbf{p}_{1cm}|^2} |\mathcal{M}|^2,\qquad(1.19)$$

where the Mandelstam invariants are $s = (p_1 + p_2)^2$, $t = (p_1 - p_3)^2$ in terms of the 4-momenta of the incoming (p_1, p_2) and outgoing (p_3, p_4) particles. The matrix element $\mathcal{M}$ is just another name for the proper vertex (1.18). If the scattering occurs at energies much larger than the mass of the particle, we can approximate it as being massless. Then $s = 4E^2$ if each particle has energy E, and $|\mathbf{p}_{1cm}|^2 = E^2$. Furthermore, the total cross section in the massless limit is given by

$$\sigma = \int\limits_{-s}^{0} dt \frac{d\sigma}{dt} = \frac{\lambda^2}{64\pi E^2}\qquad(1.20)$$

in the center of mass frame. Suppose that λ is of order unity. For particles whose energy is 100 GeV, near the limiting energy of the LEP accelerator, this is a cross section of order 5×10^{-7} GeV^{-2} $\times$ (0.197 GeV-fm)^2 = 10^{-8} fm^2 = 10^{-38} m^2 = 10^{-10} barn = 100 pb. For comparison, the cross section for $e^+e^- \to Z$ at the Z resonance, measured by LEP, is 30 nb, some 300 times larger.

Chapter 2
The Loop Expansion

In the previous example of the 4-point function, the first connected diagram arose at linear order in λ. However, if we consider the 2-point function, we already have a connected piece at zeroth order: it is simply the propagator, $\widetilde{G}_2 = i/(p^2 - m^2)$. When we compute the first order correction to this, we get the diagram of Fig. 2.1. There are 4 ways of doing the first contraction and 3 of the second, so the statistical factor is $4 \cdot 3/4! = 1/2$. In position space, this diagram is a correction to the 2-point function given by

$$\delta G_2(x, y) = -\frac{i\lambda}{2} \int d^4x' G_2(x', x') G_2(x, x') G_2(y, x'). \qquad (2.1)$$

When we transform to momentum space and remove the overall factor of $(2\pi)^4 \delta^{(4)}$ $(p_1 - p_2)$ for 4-momentum conservation, we are left with

$$\delta\widetilde{G}_2 = -\frac{i\lambda}{2} G_2(x', x')(\widetilde{G}_2)^2 = -\frac{i\lambda}{2}(\widetilde{G}_2)^2 \int \frac{d^4p}{(2\pi)^4} \frac{i}{p^2 - m^2 + i\varepsilon}. \qquad (2.2)$$

What is the physical meaning of this correction? This is most easily seen by expanding the following expression in δm^2:

$$\frac{i}{p^2 - m^2 - \delta m^2} = \frac{i}{p^2 - m^2} + \frac{i\delta m^2}{(p^2 - m^2)^2} + \cdots. \qquad (2.3)$$

This expansion has exactly the same form as $\widetilde{G}_2 + \delta\widetilde{G}_2$. We can therefore see that the loop diagram of Fig. 2.1 is nothing more nor less than a correction to the mass of the particle:

$$m_{\text{phys}}^2 = m^2 + \delta m^2; \quad \delta m^2 = i\frac{\lambda}{2} \int \frac{d^4p}{(2\pi)^4} \frac{1}{p^2 - m^2 + i\varepsilon}. \qquad (2.4)$$

J. M. Cline, *Advanced Concepts in Quantum Field Theory*,
SpringerBriefs in Physics, https://doi.org/10.1007/978-3-030-56168-0_2

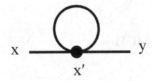

x — x' — y

Fig. 2.1 Correction to the 2-point function

Fig. 2.2 A series of corrections to the 2-point function

Fig. 2.3 A two-loop
correction to the 2-point
function

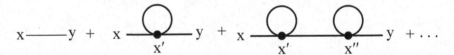

This interpretation is not merely a consequence of expanding to first order in λ. We could continue the procedure of expanding Eq. (2.3) to arbitrary orders. It corresponds to the series of diagrams in Fig. 2.2. This is not to say that Fig. 2.2 is the complete answer for the corrections to the mass. There are other corrections starting at two loops which come in addition to the one we have calculated. Notice that the extra diagrams in Fig. 2.2 relative to Fig. 2.1 have no effect on our expression (2.4) for the mass shift. What we have computed is the one-loop contribution to the mass of the particle. The extra diagrams in Fig. 2.2 which contain two or more loops are just an iteration of our one-loop result when we expand (2.3) to higher order in δm^2. An example of a higher order diagram which gives an intrinsically two-loop contribution to the mass is Fig. 2.3, called the "setting sun" diagram.

At some point above, I started using the units $\hbar = 1$. However, it is enlightening to restore the $\hbar$'s for a moment to see how they enter the loop expansion. Recall that each power of λ is accompanied by $1/\hbar$. On the other hand, each propagator, since it is like the inverse of the action, comes with $\hbar$ in the numerator. The series for the propagator can be written as

$$\frac{i\hbar}{p^2 - m^2} + \frac{i\hbar^2 \delta m^2}{(p^2 - m^2)^2} + \cdots \qquad (2.5)$$

We see that the tree diagram comes with one power of $\hbar$, the one-loop diagram has $\hbar^2$, etc. If we were to imagine taking $\hbar \to 0$, the tree diagram would be the leading contribution. It is easy to see that the same is true not just for two-point functions, but for any Green's function. The important conclusion is that *tree diagrams represent*

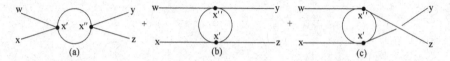

Fig. 2.4 One-loop corrections to the 4-point function

the classical contributions to a given process, while loops are quantum corrections. We can check this on another example, the 4-point function, whose first loop correction is shown in Fig. 2.4. Let us now truncate the external legs for simplicity, à la LSZ:

$$\Gamma_4 = -i\frac{\lambda}{\hbar} + O\left(\frac{\lambda^2}{\hbar^2}\left(\frac{\hbar}{p^2 - m^2}\right)^2\right). \tag{2.6}$$

Again, the loop contribution is suppressed by an additional power of $\hbar$ relative to the tree diagram.

Chapter 3
The Feynman Rules

While we are discussing the diagram of Fig. 2.4, we might as well work out its value. We have to expand the interaction in e^{iS} to second order, so that

$$\delta G_4 = \frac{1}{Z[0]} \int \mathcal{D}\phi\, e^{iS[0]/\hbar} \phi(w)\phi(x)\phi(y)\phi(z) \frac{1}{2} \left[\int d^4x' \left(-\frac{i}{\hbar}\frac{\lambda}{4!}\phi^4\right)\right]^2. \quad (3.1)$$

$$= \frac{1}{Z[0]} \int \mathcal{D}\phi\, e^{iS[0]/\hbar} \phi(w)\phi(x)\phi(y)\phi(z) \frac{1}{2} \left[\int d^4x' \left(-\frac{i}{\hbar}\frac{\lambda}{4!}\phi^4(x')\right)\right]$$

$$\left[\int d^4x'' \left(-\frac{i}{\hbar}\frac{\lambda}{4!}\phi^4(x'')\right)\right].$$

Notice that there are three topologically distinct ways of contracting the external legs with the vertices. This will be more clear when we go to momentum space (Fig. 3.1). The distinction occurs because I have already decided that positions w and x will correspond to the incoming particles, and y and z will be for the outgoing ones. Let's first treat diagram Fig. 2.4a; the other two will then follow in a straightforward way.

First the statistical factor: To connect $\phi(w)$, we can choose either the x' or the x'' vertex. That gives a factor of 2. Let's choose x'. We have 4 choices of contraction here. Next connect $\phi(x)$ to one of the remaining legs on the x' vertex; this is a factor of 3. We know that $\phi(y)$ must go with the x'' vertex, and there are 4 ways of doing this. That leaves 3 ways of connected $\phi(z)$. Finally, there are 2 ways of connecting the remaining legs between the vertices. So the statistical factor is $(2 \cdot 3 \cdot 4/4!)^2/2 = 1/2$. In position space, we have

$$\delta G_4^{(a)} = \frac{(-i\lambda)^2}{2} \int d^4x' \int d^4x''\, G_2(w, x') G_2(x, x') G_2(x', x'') G_2(x', x'') G_2(x'', y) G_2(x'', z).$$

$$(3.2)$$

Next we want to Fourier transform. Suppose the momentum going into and out of the graph is $q = p_1 + p_2$. There is also momentum going around the loop, p. Momentum is conserved at each vertex, as shown in Fig. 3.1, so one internal line

© The Author(s), under exclusive license to Springer Nature Switzerland AG 2020
J. M. Cline, *Advanced Concepts in Quantum Field Theory*,
SpringerBriefs in Physics, https://doi.org/10.1007/978-3-030-56168-0_3

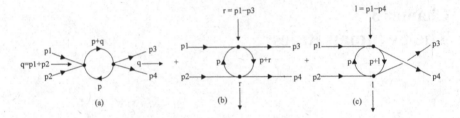

Fig. 3.1 Momentum space version of Fig. 2.4

carries momentum p, the other has $p + q$. When we Fourier transform the external legs, it just undoes the momentum integrals for those propagators, and replaces the $1/(q_i^2 - m^2)$ factors with $1/(p_i^2 - m^2)$, where q_i is the integration variable appearing in the position space propagator, and p_i is the external momentum. For the two internal progators, one of the momentum integrals is for overall 4-momentum conservation. The remaining one is the only one that is left. When we remove the external legs and the overall delta function, we get

$$\delta\Gamma_4^{(a)}(q) = \frac{(-i\lambda)^2}{2} \int \frac{d^4p}{(2\pi)^4} \frac{i}{p^2 - m^2 + i\varepsilon} \frac{i}{(p+q)^2 - m^2 + i\varepsilon}. \tag{3.3}$$

The contributions (b) and (c) are the same but with external momenta r and l replacing q. The three respective graphs are called the s-, t- and u-channels, after the Mandelstam invariants $s = q^2$, $t = r^2$ and $u = l^2$.

After going through this procedure enough times, one comes to realize that there is no need to start in position space and then Fourier transform. One can write down directly the expression for the truncated loop diagram by associating the proper factors with each element:

- for each vertex a factor $-i\lambda$;
- for each internal line, a propagator $\dfrac{i}{p^2 - m^2 + i\varepsilon}$;
- for each closed loop, a momentum integral $\int \dfrac{d^4p}{(2\pi)^4}$

In addition, one should

- conserve momentum at each vertex
- compute the statistical factor

These are the *Feynman rules* for the $\lambda\phi^4$ theory, which make it relatively easy to set up the computation for a given loop diagram. Every theory (QED, QCD, etc.) will have its own Feynman rules corresponding to the detailed way in which the particles couple to each other.

Chapter 4
Evaluation of Diagrams; Regularization

We now have formal expressions for the loop diagrams, but we still have to evaluate them. Let's start with the easiest diagram, Fig. 2.1, Eq. (2.4). The first thing we have to do is to make sense out of the pole in the propagator. This is what the $i\varepsilon$ is for. The procedure is called the *Wick rotation*. We will do the integral over p_0 separately, and use the techniques of contour integration. Notice that the poles in the complex p_0 plane are located at

$$p_0 = \pm(\sqrt{\mathbf{p}^2 + m^2} - i\varepsilon). \tag{4.1}$$

Their positions are indicated by the x's in Fig. 4.1. According to the theory of complex variables, we can deform the integration contour without changing the value of the integral as long as we don't move it though the poles. This defines the sense in which we are allowed to rotate the integration contour from being along the real p_0 axis to the imaginary one, as shown by the arrow. When we do the Wick rotation, we are making the replacement

$$p_0 \rightarrow ip_0 \tag{4.2}$$

everywhere in the integral. This is equivalent to going from Minkowski to Euclidean space, so we can also denote the new momentum variable by p_E, for the Euclidean momentum. Notice that because I am using the mostly $-$ convention for the Minkowski space metric, the square of the 4-momentum vector transforms as

$$p^2 \rightarrow -p_E^2 \tag{4.3}$$

when Euclideanizing. We thus obtain

$$\delta m^2 = -i^2 \frac{\lambda}{2} \int \frac{d^4 p_E}{(2\pi)^4} \frac{1}{p_E^2 + m^2}. \tag{4.4}$$

© The Author(s), under exclusive license to Springer Nature Switzerland AG 2020
J. M. Cline, *Advanced Concepts in Quantum Field Theory*,
SpringerBriefs in Physics, https://doi.org/10.1007/978-3-030-56168-0_4

Fig. 4.1 Integration
contours for the Wick
rotation

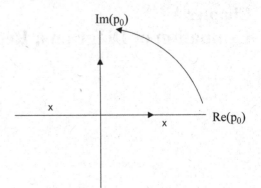

Notice that the $i\varepsilon$ prescription is no longer necessary once we have Wick-rotated. Also, the factor of i coming from the integration measure is just what we needed to cancel the explicit one that was already present, so that our mass shift is real-valued.

Now we have made sense out of the pole in the propagator, so the integral is looking a bit nicer, but there is just one problem: it diverges quadratically as $p \to \infty$! This is one of the famous ultraviolet divergences of quantum field theory, which comes from the fact that we have infinitely many degrees of freedom: virtual pairs of ϕ particles with arbitrarily large momentum are contributing to the quantum correction to the mass. We have to cut off this divergence somehow.

We don't really believe this theory is a correct description of nature up to arbitrarily high energies and momenta. Let us suppose we only think it is true up to some very high scale Λ. Then it makes sense to cut off the integral at this scale. This gives us a way of defining the correction. The integral can be factored into angular and radial parts by going to spherical coordinates in 4 dimensions:

$$
\begin{aligned}
p_1 &= p \sin\theta_1 \sin\theta_2 \cos\phi \\
p_2 &= p \sin\theta_1 \sin\theta_2 \sin\phi \\
p_3 &= p \sin\theta_1 \cos\theta_2 \\
p_4 &= p \cos\theta_1.
\end{aligned}
\tag{4.5}
$$

Then we get

$$
\delta m^2 = \frac{\lambda}{2(2\pi)^4} \int d\Omega_3 \int_0^\Lambda dp_E\, p_E^3 \frac{1}{p_E^2 + m^2},
\tag{4.6}
$$

where the volume of the 3-sphere is given by

$$
\int d\Omega_3 = \int_0^\pi \sin^2\theta_2 d\theta_2 \int_0^\pi \sin\theta_1 d\theta_1 \int_0^{2\pi} d\phi = 2\pi^2.
\tag{4.7}
$$

The radial integral can be rewritten using $u = p_E^2$ as

$$\frac{1}{2} \int_0^{\Lambda^2} du \frac{u}{u + m^2} = \frac{1}{2} \int_{m^2}^{\Lambda^2 + m^2} du \frac{(u - m^2)}{u} \tag{4.8}$$

$$= \frac{1}{2} \left(u - m^2 \ln u \, \big|_{m^2}^{\Lambda^2 + m^2} \right) \tag{4.9}$$

$$= \frac{1}{2} \left(\Lambda^2 - m^2 \ln(1 + \Lambda^2/m^2) \right), \tag{4.10}$$

so that finally

$$\delta m^2 = \frac{\lambda}{32\pi^2} \left(\Lambda^2 - m^2 \ln(1 + \Lambda^2/m^2) \right). \tag{4.11}$$

The procedure of rendering the Feynman diagram finite is called *regularization*. There are many ways of doing it; we have used the *momentum space cutoff* method here. A more elegant but less physically transparent method, dimensional regularization, will be introduced a little later on.

The result is that the mass which is physically measured is only approximately given by the original parameter m^2 which was in the Lagrangian. We refer to this parameter as the *bare* mass and we rename it m_0^2 to emphasize that it is not yet corrected by perturbation theory. The physical mass is $m_0^2 + \delta m^2$, to linear order of accuracy in perturbation theory.

Chapter 5
Renormalization

The physical mass is something that is experimentally measured, whereas the cutoff is an arbitrary parameter which is not very well defined. It might by 10 TeV or it might be 10^{19} GeV. The experimentally measured values must not depend on our arbitrary choice of Λ. This tells us that the bare parameter m_0^2 must really be a function of Λ. Defining $\hat{\lambda} = \lambda/(32\pi^2)$,

$$m_0^2(\Lambda) = m_r^2 - \hat{\lambda}\left(\Lambda^2 - m_0^2\ln(\Lambda^2/\mu^2)\right)$$

$$\longrightarrow \quad m_0^2(\Lambda) = \frac{m_r^2 - \hat{\lambda}\Lambda^2}{1 - \hat{\lambda}\ln\Lambda^2/\mu^2} \cong m_r^2(1 + \hat{\lambda}\ln\Lambda^2/\mu^2) - \hat{\lambda}\Lambda^2, \quad (5.1)$$

where m_r and μ are parameters which do not depend on Λ. The second equality in (5.1) takes account of the fact that we are only working to first order in λ. Now when we combine the bare mass and the correction to get the physical value, we obtain

$$m_{\text{phys}}^2 = m_r^2(1 + \hat{\lambda}\ln m_r^2/\mu^2), \quad (5.2)$$

which does not depend on the cutoff. The terms we added to m_0^2 which diverge as $\Lambda \to \infty$ are called *counterterms*. If the original Lagrangian was naively considered to have m_r^2 as the mass parameter, then the one-loop counterterms are what we need to add to it so that physical mass is finite at one loop order. Of course, we should continue the process to higher orders in λ and higher loops to be more accurate.

The procedure we have carried out here is known as *renormalization*. If m_0^2 was the bare mass, then $m_0^2 + \delta m^2$ is the renormalized one, at least up to finite corrections. So far it looks like this business has been somewhat pointless. We did all this work to find corrections to the mass, yet it did not allow us to predict anything, since in the end we simply fixed the parameters of the theory by comparing to experiment. However, when we continue along these lines for all the other relevant amplitudes, we will see that renormalization allows us to make very detailed predictions relating

© The Author(s), under exclusive license to Springer Nature Switzerland AG 2020

J. M. Cline, *Advanced Concepts in Quantum Field Theory*,

SpringerBriefs in Physics, https://doi.org/10.1007/978-3-030-56168-0_5

masses and scattering cross sections, and telling us how these quantities scale as we change the energy scale at which they are being probed. To appreciate this, we should next turn to the 4-point function, which controls 2-body scattering.

Before doing that however, we can already draw a very interesting conclusion from our computation of the mass. Suppose that we were talking about the Higgs boson, even though this isn't exactly the right theory for it. It is known that the Higgs mass is close to or above about 115 GeV, since LEP II finished running with a suggestive but not overwhelmingly convincing hint of a signal at that mass. On the other hand, our theory might be cut off at a scale many orders of magnitude greater, perhaps even the Planck mass at $\sim 10^{19}$ GeV. If this is the case, then there had to be a very mysterious conspiracy between the bare parameters and the quantum corrections: they had to cancel each other to a part in 10^{17} or so. This does not look natural at all, leading us to believe that there must be some physical mechanism at work. This mystery is known as the *gauge hierarchy problem*. Either there is some approximate symmetry like supersymmetry which explains why the quantum corrections aren't really as large as our theory predicts, or else there is some other new physics which invalidates our theory at energy scales which are not that much greater than the 100 GeV scale where we are presuming that it works.

Now to continue our discussion of renormalization to the 4-point function, recall that we computed the scattering cross section at tree level in Eq. (1.20), which is simply related to the tree level value of the 4-point proper vertex, $\Gamma_4 = -i\lambda$. To correct this, we should add the one-loop contribution $\delta\Gamma_4$ from Eq. (3.3). Evaluating this kind of integral is harder, and Feynman has invented a very useful trick for helping to do so. The trick is to combine the two denominator factors into a single denominator, using the identity[1]

$$\frac{1}{AB} = \int\limits_0^1 dx\, \frac{1}{(xA + (1-x)B)^2}.$$

(5.4)

The integral over x is known as the Feynman parameter integral. When we apply this to (3.3), it becomes

---

[1] The more general expression, useful for more complicated diagrams, is

$$\prod_{i=1}^k \frac{1}{D_i^{a_i}} = \frac{\Gamma(\sum_i^k a_i)}{\prod_{i=1}^k \Gamma(a_i)} \int_0^1 dx_1 \cdots \int\limits_0^1 dx_k\, \frac{\delta\left(1 - \sum_i^k x_i\right)\prod_{i=1}^k x_i^{a_i-1}}{\left(\sum_i^k D_i x_i\right)^{\sum_i^k a_i}}$$

(5.3)

in terms of gamma functions.

$$\delta\Gamma_4^{(a)}(q) = \frac{\lambda^2}{2} \int \frac{d^4 p}{(2\pi)^4} \frac{1}{(p^2 + 2xp \cdot q + xq^2 - m^2 + i\varepsilon)^2}$$

$$= \frac{\lambda^2}{2} \int \frac{d^4 p}{(2\pi)^4} \frac{1}{((p + xq)^2 + x(1-x)q^2 - m^2 + i\varepsilon)^2}$$

$$= \frac{\lambda^2}{2} \int \frac{d^4 p}{(2\pi)^4} \frac{1}{(p^2 + x(1-x)q^2 - m^2 + i\varepsilon)^2} \tag{5.5}$$

$$= i\frac{\lambda^2}{2} \int \frac{d^4 p_E}{(2\pi)^4} \frac{1}{\left(p_E^2 - x(1-x)q^2 + m^2\right)^2}, \tag{5.6}$$

and the crossed diagrams (b,c) are the same but with $q \to r$ and $q \to l$. In (5.5) we assumed that it is permissible to shift the variable of integration, and in (5.6) we did the Wick rotation on p, but not on q—the latter is still a Minkowski 4-momentum. Notice that the shift of integration variable might be considered somewhat of a cheat, if we were thinking that the original p variable was cut off at Λ. That cutoff will get shifted when we change variables. However, this only introduces an error of order q^2/Λ^2, which vanishes in the limit $\Lambda \to \infty$. Now we can do the integral as we did before. Defining $M^2(q) = m^2 - x(1-x)q^2$,

$$\delta\Gamma_4^{(a)}(q) = i\frac{\lambda^2}{32\pi^2} \int_0^1 dx \int_0^{\Lambda^2} du \frac{u}{(u + M^2(q))^2} \tag{5.7}$$

$$= i\frac{\lambda^2}{32\pi^2} \int_0^1 dx \int_{M^2(q)}^{\Lambda^2 + M^2(q)} du \frac{u - M^2(q)}{u^2} \tag{5.8}$$

$$= i\frac{\lambda^2}{32\pi^2} \int_0^1 dx \left[\ln(1 + \Lambda^2/M^2(q)) - (1 + M^2(q)/\Lambda^2)^{-1} \right] \tag{5.9}$$

$$\cong i\frac{\lambda^2}{32\pi^2} \int_0^1 dx \left[\ln(\Lambda^2/M^2(q)) - 1 \right]. \tag{5.10}$$

This result is much richer than our computation of the mass shift: here we have found the the quartic coupling gets renormalized not just by a constant shift, but in fact a function of the external momentum. Let's defer the study of this finite part for later, since we are discussing the renormalization of the theory. First, notice that this result is less divergent than the mass shift; this is only log divergent rather than quadratic. Second, there is no momentum dependence in the divergent part, which is important for renormalizability. The divergence can be removed by defining the *bare coupling* λ_0 as

$$\lambda_0 = \lambda_r + \frac{3\lambda_r^2}{16\pi^2} \ln \Lambda/\mu, \tag{5.11}$$

where, again, λ_r and μ are independent of Λ. (The factor of 3 comes from adding the crossed contributions.) If we had found that the nontrivial function of external momenta was multiplied by a divergent coefficient, we would be in trouble. In that case we would have to modify the original (bare) theory with a counterterm that did not resemble the simple operators in the action we started with. In fact, a theory with a nontrivial function of momentum in the action would be *nonlocal*. If you Fourier transform a generic function f of momentum back to position space, you would generally need arbitrarily large numbers of derivatives. For example, by Taylor-expanding f,

$$f(q_1 + q_2)\tilde{\phi}(q_1)\tilde{\phi}(q_2)\tilde{\phi}(q_3)\tilde{\phi}(q_4) = \sum_n f_n(q_1 + q_2)^{2n}\tilde{\phi}_1\tilde{\phi}_2\tilde{\phi}_3\tilde{\phi}_4 \quad (5.12)$$

$$\rightarrow \sum_n f_n(\partial_\mu \partial^\mu)^n(\phi_1\phi_2)\phi_3\phi_4, \quad (5.13)$$

which would be a rather ugly term to add to a Lagrangian. We say that an interaction with arbitrarily large number of derivatives is nonlocal because we know that

$$\phi(x)\phi(x + a) = \phi(x)e^{a\frac{d}{dx}}\phi(x) \quad (5.14)$$

$$= \phi(x) \sum_n \frac{1}{n!}\left(a\frac{d}{dx}\right)^n \phi(x), \quad (5.15)$$

for example. This would be something like a theory with instantaneous action at a distance (the distance being a). Such an interaction contradicts our cherished belief that all interactions are fundamentally local in nature, occuring at a specific point in spacetime, at least down to the $(100 \text{ GeV})^{-1}$ distance scales which have been experimentally explored.

Fortunately, we are able to absorb the new infinity simply into the coupling constant λ. In this way, the scattering cross section computed at one loop is independent of Λ, just like the mass shift was. Similarly to that example, we must determine an unknown combination $\lambda_r - \frac{\lambda_r^2}{16\pi^2} \ln \mu$ by comparing the theory to the measured value of the cross section.

Again, we come to the question: where is the predictive power of the theory if we are merely fixing the unknown parameters in the Lagrangian by comparing to measured quantities? The answer is that by fixing just these *two* quantities, the mass and the coupling, we are able to compute everything else, and thus make predictions for an arbitrarily large number of other physical processes. This is the crux of *renormalizability*: a theory is said to be renormalizable if only a finite number of parameters need to be shifted in order to remove all the infinities.

For example, we could consider an inelastic scattering process, in which two particles collide to create four particles. Some of the tree diagrams for this process are shown in Fig. 5.1. Now consider a one-loop correction, Fig. 5.2. For simplicity, let's assume that the net external momentum entering and leaving the graph is zero. This assumption can't be true for a real process in which all the external particles are

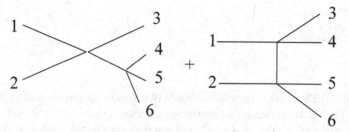

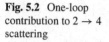

Fig. 5.1 Tree-level contributions to $2 \to 4$ scattering

Fig. 5.2 One-loop contribution to $2 \to 4$ scattering

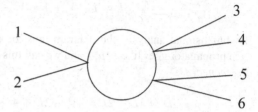

on their mass shells (*on-shell*, meaning each particle satisfies $p^2 = m^2$), since in that case we should have at least $4\,m$ of energy entering the diagram in order to produce four particles at rest. However, the point I want to make concerns the contribution to this diagram from very high-momentum virtual particles, which is insensitive to the small external momentum. The Feynman rules tell us that (after Wick-rotating)

$$\Gamma_6 \sim \int \frac{d^4 p_E}{(2\pi)^4} \frac{1}{(p_E^2 + m^2)^3},\tag{5.16}$$

which is a convergent integral. Similarly, it is easy to see that higher-point functions are even more convergent. At one-loop order, the only divergent quantities are Γ_2 and Γ_4, and this is why we can renormalize the theory using only two parameters, m^2 and λ.

The estimate (5.16) is an example which leads the way to a general *power-counting* argument of the type which was first introduced by S. Weinberg, for establishing the renormalizability of a theory. The argument goes like this: let's consider an arbitrarily complicated Feynman diagram, which has L loops, V vertices, and E external legs. We would like to know which diagrams in the theory are divergent, to extend the concept of renormalization beyond just one loop. We know that each loop will give us an integral of the form $\int \frac{d^4 p}{(2\pi)^4}$; how badly these integrals can diverge in the UV (ultraviolet, high momentum) depends on how many propagators there are. To count the number of propagators, we need the number of internal lines, I. Before doing contractions, there are $4V - E$ lines available to contract, and afterwards there are

$$I = 2V - E/2\tag{5.17}$$

Fig. 5.3 "Figure eight"
vacuum diagram, which is
quartically divergent

internal lines. There is also a relationship between these quantities and the number
of loops. Each internal line has a momentum associated with it, and each vertex must
conserve momentum. The total number of internal lines minus vertices is related to
the number of independent momenta, which is also the number of loops:

$$L = I - V + 1 = V - E/2 + 1. \tag{5.18}$$

By dimensional analysis, the diagram must be divergent if $4L$ exceeds $2I$ (since
each propagator goes like $1/p^2$), and we call this difference the *superficial degree of
divergence*, D:

$$D = 4L - 2I = 4V - 2E + 4 - 4V + E = 4 - E. \tag{5.19}$$

D is the minimum power of Λ with which a diagram with E external legs can be
expected to scale: $E = 2$ gives the quadratic divergence $D = 2$, and $E = 4$, $D = 0$
is log divergent. We did not yet discuss the vacuum diagrams, with no vertices, as in
Fig. 5.3, but it is easy to see that they also obey this rule, being quartically divergent.
(Vacuum diagrams give contributions to the energy density of the vacuum, *i.e.*, the
cosmological constant.)

According to this superficial counting, one might expect all the higher point functions to be finite, as indeed our one-loop examples proved to be. However, one
can always embed one of the divergences from Γ_2 or Γ_4 inside a diagram with an
arbitrary number of external legs, rendering it divergent, which is why this kind of
power-counting criterion *is* superficial. For example, insertion of a mass correction
on the external legs of the 6-point function produces a quadratic divergence, while
the other diagram shown (b) is log divergent (Fig. 5.4). However, it should not be
too surprising that these divergences, which are not accounted for by the simple
power-counting argument, do not pose any new problems—in fact, they are already
automatically taken care of by the renormalization of the mass and coupling which
we have already done!

For example, Fig. 5.4a corresponds to correcting the mass of one of the external
legs. Really, we should correct the masses of all the external particles, not just one
of them. Mass renormalization is somewhat special in this respect. To correct the
propagator, we had to imagine summing up the entire geometric series of diagrams
which were of the form of Fig. 2.1. As the great theorist and pedagogue of field
theory Sidney Coleman put it, this is the one nonperturbative step you must always
do in perturbation theory. In practical terms, it means that we don't really have to
write down the diagrams like Fig. 5.4a; we merely keep in mind that masses will get
renormalized.

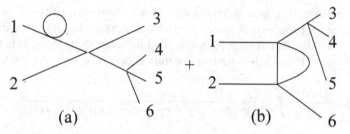

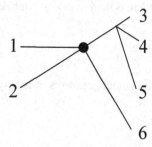

Fig. 5.4 Divergent contributions to 6-point function

Fig. 5.5 Divergent counterterm contribution to 6-point function, which cancels the divergent part of Fig. 5.4b

What about the log divergence in Fig. 5.4b? It is pretty obvious that this gets canceled by the counterterm (divergent part) from the corresponding vertex of the tree diagram shown in Fig. 5.5.

We can make a distinction between the kinds of diagrams which are intrinsically important for renormalization, versus those for which renormalization automatically works because of taking care of the first kind. Notice that the diagrams in Fig. 5.4 can be chopped up into smaller diagrams by severing just a single line. These are call *one-particle reducible* graphs, abbreviated 1PR. On the other hand, the diagrams like Figs. 2.1, 2.3, 2.4 cannot be split into two separate graphs by cutting a single line. They are called *one-particle irreducible*, 1PI. The latter are the important ones from the point of view of renormalization; they are the *primitively divergent* diagrams, if they are divergent. Any other divergences that occur can be considered to come from the embedding of the primitively divergent diagrams inside a more complicated diagram.

So far we have concentrated mainly on one-loop effects. Does anything qualitatively new happen at two loops? In fact, yes. Notice that our counterterms so far only include two of the three operators of the original Lagrangian:

$$\mathcal{L}_{\text{c.t.}} = -\frac{1}{2}\hat{\lambda}(m_r^2 \ln \Lambda^2/\mu^2 - \Lambda^2)\phi^2 - \frac{3\lambda_r^2}{4! \, 16\pi^2} \ln(\Lambda/\mu)\phi^4 \qquad (5.20)$$

These are the divergent parts of the bare Lagrangian. But we might have also expected to obtain a term of the form

$$Z_\phi(\Lambda)\partial_\mu\phi\partial^\mu\phi \qquad (5.21)$$

where Z_ϕ is some divergent function of Λ. This is the normal situation, but in the present theory, it happens to first arise at two loops, through the setting sun diagram, Fig. 2.3. The thing that distinguishes this from the mass shift, Fig. 2.1, is that the divergent part of Fig. 2.3 does have momentum dependence. It gives a contribution

$$\delta\Gamma_2(q) = \frac{(-i\lambda)^2}{6} \int \frac{d^4p}{(2\pi)^4} \frac{d^4r}{(2\pi)^4} \frac{i}{((p-r)^2 - m^2)} \frac{i}{((p+q)^2 - m^2)} \frac{i}{(r^2 - m^2)} \tag{5.22}$$

which we can Taylor-expand in powers of the external momentum q. Notice that odd powers of q must have vanishing coefficients because $\delta\Gamma_2(q)$ is a Lorentz scalar, and there is no 4-vector to contract with other than q_μ. Let us write

$$\delta\Gamma_2(q) = \sum_{n=1} q^{2n} \delta\Gamma_2^{(n)}(\Lambda) \tag{5.23}$$

By counting powers, it is easy to see that the only divergent terms in this expansion are

$$\delta\Gamma_2^{(0)} = \delta\Gamma_2(0) \sim \Lambda^2 \tag{5.24}$$

$$\delta\Gamma_2^{(1)} \sim \ln\Lambda \tag{5.25}$$

The first term, with no momentum dependence, is a new, two-loop contribution to the mass correction. The second term, since it goes like q^2, is the Fourier transform of the operator $\partial_\mu\phi\partial^\mu\phi$. All the higher terms in the expansion are convergent without the cutoff and do not require renormalization. Let us suppose that $\delta\Gamma_2^{(1)}$ has the form

$$\delta\Gamma_2^{(1)} = a + b\ln\Lambda \tag{5.26}$$

If we add a term

$$\mathcal{L}_{c.t.} = \text{previous terms} + \frac{1}{2}\delta Z_\phi \, \partial_\mu\phi\partial^\mu\phi \tag{5.27}$$

to the counterterm Lagrangian, with

$$\delta Z_\phi = -(b\ln\Lambda/\mu) \tag{5.28}$$

for some μ, then the divergent part of the new contribution will be canceled. Let's verify this by treating $\mathcal{L}_{c.t.}$ as a perturbation to the 2-point function. In the path integral, it gives

$$\int \mathcal{D}\phi \, e^{iS} \phi(x)\phi(y) \frac{i}{2} \int d^4x' \, \delta Z_\phi \, \partial_\mu\phi\partial^\mu\phi \tag{5.29}$$

which contributes

$$\delta\Gamma_2(q) = q^2 \delta Z_\phi + \dots \tag{5.30}$$

Fig. 5.6 Momentum-space
version of setting sun
diagram

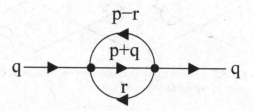

to $\delta\Gamma_2(q)$. The factor of 1/2 is canceled by the number of ways of doing the contractions. This result therefore cancels the divergent part of Fig. 5.6.

This last kind of correction is called *wave function renormalization*, because it can be absorbed into a rescaling of the field. Indeed, the renormalized Lagrangian has the form

$$\mathcal{L}_{\text{ren}} = \frac{1}{2}\left(1 + \delta Z_\phi(\Lambda)\right)\partial_\mu\phi\partial^\mu\phi - \frac{1}{2}\tilde{m}_0^2(\Lambda)\phi^2 - \frac{\tilde{\lambda}_0(\Lambda)}{4!}\phi^4. \tag{5.31}$$

I have used tildes on $\tilde{m}_0^2$ and $\tilde{\lambda}_0$ to distinguish them from the final renormalized couplings which we will now define. If we rescale the field by $\phi = \phi_0/\sqrt{Z_\phi}$, where $Z_\phi = 1 + \delta Z_\phi$, then

$$\mathcal{L}_{\text{ren}} = \frac{1}{2}\partial_\mu\phi_0\partial^\mu\phi_0 - \frac{1}{2}\frac{\tilde{m}_0^2}{Z_\phi}(\Lambda)\phi_0^2 - \frac{1}{4!}\frac{\tilde{\lambda}_0}{Z_\phi^2}(\Lambda)\phi_0^4. \tag{5.32}$$

In light of this, we should redefine the bare mass and coupling to include the effect of wave function renormalization:

$$\mathcal{L}_{\text{ren}} \to \frac{1}{2}\partial_\mu\phi_0\partial^\mu\phi_0 - \frac{1}{2}m_0^2(\Lambda)\phi_0^2 - \frac{\lambda_0}{4!}(\Lambda)\phi_0^4, \tag{5.33}$$

where

$$m_0^2 = \frac{\tilde{m}_0^2}{Z_\phi}; \qquad \lambda_0 = \frac{\tilde{\lambda}_0}{Z_\phi^2}. \tag{5.34}$$

The final form (5.33) is how the renormalized Lagrangian is conventionally defined: the kinetic term of the bare field is canonically normalized. The divergences appearing in the newly defined counterterms for the mass and the coupling no longer exactly cancel the divergences of the diagrams like Figs. 2.1 and 2.4 and their higher-loop generalizations; they also include the effects of wave function renormalization.

For example, the divergent δZ terms we have absorbed into λ are such as to partially cancel out the corresponding divergences in Fig. 5.7. However, we truncate the external lines, including the setting sun diagrams, when we implement the LSZ procedure, so the extra factors of $1 + \delta Z$ in λ_0 have to be taken out again to get a finite 1PI amplitude if we are computing it from the renormalized Lagrangian (5.33).

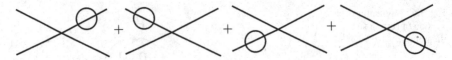

Fig. 5.7 Two-loop divergent contributions to the 1PR Green's functions, which are partially canceled by the δZ factors hidden in λ_0

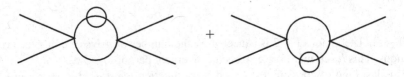

Fig. 5.8 Three-loop divergent contributions to the 1PR Green's functions, which are only partially
canceled by the δZ factors hidden in λ_0

For the 4-point function, we write the renormalized amplitude (which is finite) in
terms of the bare one (the one computed from (5.33)),

$$\Gamma_4^{(r)} = \sqrt{Z_\phi}^4 \Gamma_4^{(0)} \tag{5.35}$$

$$= (1 + \delta Z_\phi)^2 \Gamma_4^{(0)} \tag{5.36}$$

$$\cong (1 + 2\delta Z_\phi)\Gamma_4^{(0)}, \tag{5.37}$$

where we expanded perturbatively in the last approximation. The factor of $2 = 4/2$
is there because there are 4 external legs in the diagram. Notice that the amplitudes in
question are the 1PI vertices. Also, in the above description, $\Gamma_4^{(0)}$ is the proper vertex
that was computed from the renormalized Lagrangian (5.33), including all the loop
corrections to the relevant order. But we call it the unrenormalized vertex since it is
not finite until w.f.r. is applied.

Another example of how wave function renormalization works is provided by
Fig. 5.8. Four of the eight $\delta Z/2$ factors hidden in the vertices cancel the divergence
associated with w.f.r. in the internal lines. However, the remaining four would like
to do the same for the external lines, which are not present in the proper vertex. So
again Eq. (5.37) must be applied.

In this discussion we have stressed the removal of divergences, but you should
keep in mind that w.f.r. would still be necessary even if δZ was completely finite.
Otherwise the amplitudes we compute would not be correctly normalized. A field
ϕ is canonically normalized if its creation and annihilation operators give the actual
number of particles in a state (the number operator) when acting upon that state:
$a^\dagger a|\Phi\rangle = N|\Phi\rangle$. Such a field has a propagator $i/(p^2 - m^2)$, not $iZ_\phi/(p^2 - m^2)$.
But in general it is necessary to start with a tree-level propagator of the latter form so
that when we add the loop corrections to it, the physical propagator has the correct
normalization.

*N.B. when you solve problem 4(c), compute the renormalized 3-point function, so
you can get some practice with implementing w.f.r.*

Let's take a moment to clarify some potentially confusing terminology. The original Lagrangian we write down before worrying about renormalization is the *bare Lagrangian*, while (5.33) is the renormalized one, even though it is written in terms of the bare parameters and the bare field. We can split $\mathcal{L}_{\text{ren}}$ into two pieces, a finite part and a counterterm part:

$$\mathcal{L}_{\text{ren}} = \mathcal{L} + \mathcal{L}_{\text{c.t.}}, \tag{5.38}$$

where $\mathcal{L}_{\text{c.t.}}$ contains all the pieces which diverge as $\Lambda \to \infty$. The parameters which appear in $\mathcal{L}$ are closely related to what we call the *renormalized parameters*, and to the physical measured values:

physical parameter = (parameter in $\mathcal{L}$) + (finite part of loop calculations) (5.39)

There is another way of specifying what we mean by the renormalized parameters. It is always possible to write the bare parameters as

$$m_0^2(\Lambda) = Z_m(\Lambda)m_r^2; \tag{5.40}$$
$$\lambda_0(\Lambda) = Z_\lambda(\Lambda)\lambda_r, \tag{5.41}$$

where m_r^2 and λ_r are finite and Λ-independent—these are the renormalized values—while all the Λ-dependence is contained in the factors Z_m and Z_λ. It is important to notice that this definition does not uniquely specify the values of the renormalized parameters, however. The reason is the ambiguity in separating divergent from convergent contributions. Since Λ has dimensions of mass, we cannot simply subtract $\ln \Lambda$; we have to subtract $\ln \Lambda/\mu$, where μ is some arbitrarily chosen mass scale. This arbitrariness will play a crucial role in defining the concept of the *renormalization group*.

It is remarkable that our renormalized Lagrangian has exactly the same form as the bare one. In fact this is the definition of a renormalizable theory: no new operators are required to be added to the theory to make it finite. This did not have to be the case by any means: renormalizable theories are a set of measure zero in the space of all possible theories. Consider the result of adding an interaction term of the form

$$\mathcal{L}_{\text{int}} = -\frac{g_n}{M^{n-4}}\phi^n \tag{5.42}$$

to the Lagrangian. We can repeat the power counting argument we did previously to find out which diagrams are superficially divergent. To be even more general, let's suppose we are working in d spacetime dimensions instead of 4. Then

$$I = nV/2 - E/2; \tag{5.43}$$
$$L = I - V + 1 = (n/2 - 1)V - E/2 + 1; \tag{5.44}$$
$$D = dL - 2I = (d(n/2 - 1) - n)V - (d/2 - 1)E + d \tag{5.45}$$
$$= (n - 4)V - E + 4 \text{ for } d = 4 \tag{5.46}$$

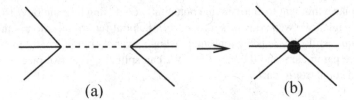

(a)   (b)

Fig. 5.9 a Virtual exchange of a heavy particle (dashed line) giving rise to **b** an effective ϕ^6 interaction

This gives us the important result that when $d = 4$ and $n > 4$, an E-point function becomes more and more divergent as we add more vertices (hence loops). Then in addition to the 0-, 2- and 4-point functions being divergent, as we had in $\lambda\phi^4$ theory, we get divergences in amplitudes with arbitrarily large numbers of external legs, provided we go to high enough order in perturbation theory. To remove the infinities from this theory, we have to add counterterms of the form (5.42) to all orders, *i.e.*, infinitely many of them. But to fix the values of the finite parts of the counterterms, we would have to match to the measured values of the cross sections for $2 \rightarrow n$ scattering processes, for all n. Hence there is no predictive power in such a theory—it requires infinitely many experimental inputs.

For this reason, nonrenormalizable theories were considered to be unphysical in the early days of quantum field theory—and even until relatively recent times. This viewpoint has changed since the crucial contributions of K. G. Wilson to the theory of renormalization. We now recognize that operators of the form (5.42) *will be* present in any realistic theory, which does not pretend to be accurate to arbitrarily high energies. The reason the new operators are not a problem is that their coefficients are small, because the mass scales which suppress them are large. For example if we have a heavy particle ψ of mass M, coupled to ϕ with the interaction $g\phi^3\psi$, then the virtual exchange of a ψ particle as shown in Fig. 5.9 produces the effective interaction

$$-\frac{g^2}{M^2}\phi^6 \tag{5.47}$$

at low external momenta, $p^2 \ll M^2$. But at large momenta, one sees that the factor of $1/M^2$ is really coming from the internal propagator,

$$-\frac{g^2}{M^2}\phi^6 \rightarrow \frac{g^2}{p^2 - M^2}\phi^6 \tag{5.48}$$

In this way, what looked like a nonrenormalizable operator at low energy can be seen to originate from a renormalizable theory, that contained an additional heavy particle. We say that we have *integrated out* the heavy particle (in the sense of having done the path integral over ψ) to obtain the effective interaction (5.47).

From (5.45) we can see which theories are renormalizable in other dimensions. In $d = 6$, ϕ^3 is the highest power that is allowed. For larger d, any n greater than 2 gives a nonrenormalizable theory. On the other hand, in $d = 2$, *all* operators are allowed.

Chapter 6
Running Couplings and the Renormalization Group

So far we have introduced renormalization merely as a trick for removing the ultra-violet divergences from a quantum field theory. However, it has much deeper implications than that. The major concept that will be revealed is that physical parameters like the coupling constant are not really constants, but they depend on the scale of energy at which the experiment is performed. The couplings are said to *run* in response to changes in the energy scale.

We can see this behavior in $\lambda\phi^4$ theory if we return to the one-loop correction to the 4-point function, which was partially calculated in Eq. (5.10). Let's finish the computation now. All that remains is to do the Feynman parameter integration. If we write $M^2(q) = m^2(1 + (4/a)(x^2 - x))$ where $a = 4m^2/q^2$, then [5]

$$\delta\Gamma_4^{(a)} = i\frac{\lambda^2}{32\pi^2}\int_0^1 dx\left[\ln(\Lambda^2/m^2) - 1 - \ln(1 + (4/a)(x^2 - x))\right] \quad (6.1)$$

$$= i\frac{\lambda^2}{32\pi^2}\left[\ln(\Lambda^2/m^2) + 1 - \sqrt{1-a}\ln\left(\frac{\sqrt{1-a}+1}{\sqrt{1-a}-1}\right)\right]. \quad (6.2)$$

For $q^2 \gg m^2$ ($a \ll 1$), this becomes

$$\delta\Gamma_4^{(a)} = i\frac{\lambda^2}{32\pi^2}\left[\ln(\Lambda^2/q^2) + 1\right] \quad (6.3)$$

and, by adding to this the counterterm, (*i.e.*, by replacing λ by λ_0), and the crossed-channel contributions, the complete 4-point function at one loop is given by

The original version of this chapter was revised: The errors in this chapter have been corrected. The correction to this chapter can be found at https://doi.org/10.1007/978-3-030-56168-0_16

J. M. Cline, *Advanced Concepts in Quantum Field Theory*,
SpringerBriefs in Physics, https://doi.org/10.1007/978-3-030-56168-0_6

$$\Gamma_4 = -i\lambda_r \left(1 + \frac{\lambda_r}{32\pi^2} \sum_{q^2=s,t,u} \ln \frac{q^2}{\mu^2} \right) \tag{6.4}$$

$$\longrightarrow \lambda_{\text{eff}}(E) \cong \lambda_r \left(1 + \frac{3\lambda_r}{32\pi^2} \ln \frac{4E^2}{\mu^2} \right). \tag{6.5}$$

The last approximation is assuming that s, t and u are of the same order of magnitude, which is true for generic scattering angles.[1]

This result has the important implication that the effective value of λ is energy-dependent, and it increases logarithmically as we go to higher energies. (Recall that in the center of mass frame, $q^2 = s = 4E^2$.) This is our first example of a running coupling, and it gives a hint about how renormalization determines the running. Notice that the dependence on energy is tracked by the dependence on the arbitrary scale μ which was introduced in the process of renormalization. This μ-dependence in turn goes with the dependence on the cutoff Λ. The divergent part of the diagram was much easier to compute than the finite part, yet it appears that we could possibly have deduced the dependence on q^2 (when $q^2 \gg m^2$) just by knowing the divergent part. This is just what the renormalization group does for us, as I shall now try to explain.

It was mentioned earlier that there is a fundamental ambiguity in choosing the counterterms because there is no unique definition of the divergent part of a diagram. This ambiguity manifests itself in the fact that the value of μ is arbitrary. Suppose we measure the cross section at some energy E_0. From this we deduce the value of $\lambda_{\text{eff}}(E_0)$. This *effective coupling* is the value of λ that gives the measured value for the cross section when we just do a tree-level calculation. However, this only fixes some combination of the two parameters λ_r and μ. We can always choose a different value of μ and compensate by shifting the value of λ_r. Since the value of μ is arbitrary, no physical quantity can depend upon it. Thus, similar to the way in which λ_0 had to be a function of Λ, λ_r must be regarded as function of μ:

$$\lambda_{\text{eff}}(E) = \lambda_r(\mu) \left(1 + \frac{3\lambda_r(\mu)}{32\pi^2} \ln \frac{4E^2}{\mu^2} \right). \tag{6.6}$$

μ is known as the *renormalization scale*. It is closely related to the *subtraction point*. The latter is another name for the value of the external momenta at which the experiment was done, that was used to fix the value of the renormalized coupling. (It is the scale at which the divergent parts of the amplitude were subtracted, hence the name.) If we choose $\mu = 2E_0$, the renormalized coupling is equal to the effective coupling, which is rather nice.

---

[1]You might worry about the fact that t is negative, so that we have the logarithm of a negative number for the t-channel contribution. This is not an intrinsically bad thing since Γ_4 is allowed to be complex. We will discuss the analytic behavior of this amplitude in more detail later on. Suffice it to say that for large $|t|$, the term in parentheses is dominated by its real part, which is gotten from taking the absolute value of the argument of the log.

A technical point: it is also permissible to choose the subtraction point to be an unphysical value of the external momenta, that is, where the vertex is off-shell. For example, it is sometimes mathematically simpler to renormalize for vanishing external momenta, $p_i = 0$. The subtraction point can be different for each amplitude Γ_n. For Γ_4 one might choose $p_i \cdot p_j = M^2(\delta_{ij} - 1/4)$ since this gives $s = t = u = M^2$. However it is done, the renormalized couplings will then be simply related to an amplitude which is analytically continued away from its physical value. One could always translate this back into a more physical subtraction scheme, since it is similar to a change in the choice of μ, upon which no physics can depend.

Now let's explore the consequences of demanding that no physical quantity depends on μ. In the present discussion, λ_{eff} has direct physical meaning, being defined in terms of the measured value of the cross section. Thus

$$\mu \frac{\partial}{\partial \mu} \lambda_{\text{eff}}(E) = 0 = \mu \frac{\partial \lambda_r}{\partial \mu} - \frac{3\lambda_r^2}{16\pi^2} + O\left(\lambda_r \mu \frac{\partial \lambda_r}{\partial \mu}\right), \tag{6.7}$$

which implies that

$$\beta(\lambda_r) \equiv \mu \frac{\partial \lambda_r}{\partial \mu} = \frac{3\lambda_r^2}{16\pi^2} + O(\lambda_r^3). \tag{6.8}$$

This is known as the *beta function* for the coupling λ, and it will play a very important role in the theory of renormalization. We can integrate $\beta(\lambda_r)$ between μ_0 and μ to find the μ-dependence of λ_r:

$$\lambda_r(\mu) = \frac{\lambda_r(\mu_0)}{1 - \frac{3\lambda_r(\mu)}{16\pi^2} \ln \frac{\mu}{\mu_0}} \tag{6.9}$$

$$= \lambda_r(\mu_0) \left(1 + \frac{3\lambda_r(\mu_0)}{16\pi^2} \ln \frac{\mu}{\mu_0}\right) + O(\lambda_r^3). \tag{6.10}$$

The expansion (6.10) is in recognition of the fact that we have not yet computed higher order terms in the perturbation series, so we cannot trust (6.9) beyond order λ_r^2 (be warned however that I am going to partially take back this statement in a moment!). We see that $\lambda_r(\mu)$ depends on μ in exactly the same way as $\lambda_{\text{eff}}(E)$ depends on E (on $2E$, to be precise—due to the fact that the total energy entering the diagram is $2E$). Thus the renormalized coupling is very closely related to the physically measured coupling, and the way it runs does tell us about the energy-dependence of the latter. We quickly become accustomed to thinking about the two quantities interchangeably. This is the real power of renormalization.

Moreover, there is a way of making the connection between λ_{eff} and λ_r stronger. Since we have the freedom to choose the value of μ at will, we can ask whether some choices are more convenient than others. Suppose we are doing measurements near some fixed energy scale, like the mass of the Z boson, $E = M_Z$. It makes sense to choose $\mu = 2E$, because then the correction of order λ_r^2 vanishes in Eq. (6.6), and we have $\lambda_{\text{eff}}(E) = \lambda_r(\mu)$. For this value of μ, we get the correct answer already from the tree-level calculation. Of course this is neglecting higher orders in perturbation theory, but it seems plausible that minimizing the size of the first order corrections will also lead to the higher order corrections being relatively small.

This point can be seen explicitly in the expansion of the integrated result (6.9). Suppose we could believe this result to all orders in λ_r, and further suppose that we had chosen the logarithm to be large so that the $O(\lambda_r^2)$ correction is sizeable compared to λ_r. Then it easily follows that all the higher order corrections are large too, since the expansion has the form $1 + x + x^2 + \cdots$. On the other hand, we wondered if such a conclusion could really be drawn, since we have not computed the $O(\lambda_r^3)$ contributions. Interestingly, the higher terms in our expansion do have meaning: they contain what are known as the *leading logarithms* in the perturbative expansion of the 4-point function. Consider the effect of the 2-loop contribution to λ_{eff} on the beta function, which could be written generically in the form

$$\beta(\lambda_r) = b_1\lambda^2 + b_2\lambda^3 + \cdots . \tag{6.11}$$

If we truncate at the cubic order and integrate, we get

$$\lambda_r(\mu) = \frac{\lambda_r(\mu_0)}{1 - b_1\lambda_r(\mu_0)\ln\frac{\mu}{\mu_0} + \frac{b_2}{b_1}\lambda_r(\mu_0)\ln\left(\frac{1/\lambda_r(\mu)+b_2/b_1}{1/\lambda_r(\mu_0)+b_2/b_1}\right)} \tag{6.12}$$

$$\cong \frac{\lambda_r(\mu_0)}{1 - b_1\lambda_r(\mu_0)\ln\frac{\mu}{\mu_0} - b_2\lambda_r^2(\mu_0)\ln\frac{\mu}{\mu_0} + \ldots} . \tag{6.13}$$

If we expand this again in powers of $\lambda_r(\mu_0)$, we get

$$\lambda_r(\mu) \cong \lambda_r(\mu_0)\left(1 + \left(b_1\lambda_r(\mu_0) + b_2\lambda_r^2(\mu_0)\right)\ln\frac{\mu}{\mu_0} + \left(b_1\lambda_r(\mu_0)\ln\frac{\mu}{\mu_0}\right)^2 + \cdots\right). \tag{6.14}$$

We see that the b_2 term is subleading to the b_1^2 term if the log is large. This is what is meant by leading and subleading logs. By integrating the beta function at order λ_r^2, we sum up all the leading logs to all orders in perturbation theory. The b_2 term gives us the next-to-leading order (NLO) logs, b_3 the next-to-next-to leading (NNLO), and so on. The existence of leading and subleading logs can also be guessed from comparing some of the higher order diagrams for the 4-point function, as in Fig. 6.1. The first of these (a) is just a product of the one loop diagram, as far as the dependence on external momentum goes, so it has two powers of logs. The second (b) can have only a single log because the loop with three propagators is UV convergent, and only the loop with the single propagator contributes a log. The miracle is that renormalization is able to tell us how diagram (a) contributes to the leading dependence on energy of λ_{eff} even though we only computed the one-loop contribution, Fig. 3.1. This process of summing up the leading logs is called the *Renormalization Group (RG) improved* perturbation expansion.

By now you must be really wondering what the Renormalization Group *is*. It is only trivially a group in the sense of group theory. An element of the renormalization group is a scale transformation of the renormalization scale [6]:

$$\mu \to e^t \mu. \tag{6.15}$$

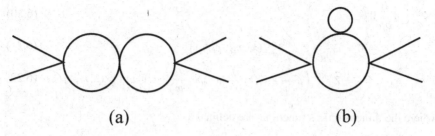

Fig. 6.1 a Leading and b subleading log contributions to the running of λ

Two successive transformations of this kind are another scale transformation,

$$e^{t_1} e^{t_2} \mu = e^{t_1 + t_2} \mu, \tag{6.16}$$

so it is obviously a group, but that is not really the point. The goal is to deduce how Green's functions scale with the external momenta, as we have illustrated in the case of the 4-point function. Let's now do it more generally, for an n-point function.

Imagine starting with the renormalized theory and computing the n-point proper vertex with a cutoff. It depends on the bare parameters, the cutoff, and the external momenta p_i: $\Gamma_n(\lambda_0, m_0, \Lambda, p_i)$. Recall that it is not enough to just use the properly defined parameters λ_0, m_0 to get finite results as $\Lambda \to \infty$; we also need to do wave function renormalization:

$$\Gamma_n^{(r)} = Z_\phi^{n/2} \Gamma_n^{(0)}. \tag{6.17}$$

$\Gamma_n^{(r)}$ is a physical observable, and therefore you might guess that it must not depend on the arbitrary renormalization scale that was introduced in the process of removing UV divergences. Strangely, however, it is $\Gamma_n^{(0)}$ which is really independent of μ. $\Gamma_n^{(0)}$ is a function of m_0^2, λ_0, and Λ, and m_0^2, λ_0 are actually independent of μ. Indeed, you can see from comparing (5.11) and (6.10) that λ_0 does not depend on μ because the implicit μ dependence in $\lambda_r(\mu)$ cancels the explicit dependence from $\ln(\Lambda/\mu)$. Similarly, the implicit μ dependence in $m_r^2(\mu)$ cancels the explicit dependence from $\ln(\Lambda/\mu)$ in Eq. (5.1). Hence we can write

$$\mu \frac{\partial}{\partial \mu} \Gamma_n^{(0)}(\lambda_0(\Lambda), m_0^2(\Lambda), \Lambda) = 0. \tag{6.18}$$

On the other hand, Z_ϕ does depend explicitly on μ, with the functional form

$$Z_\phi = 1 + c\lambda_r^2 \ln(\Lambda/\mu) \tag{6.19}$$

for some constant c, and it is clear from the way in which λ_r is known to depend on μ that the μ dependence in Z_ϕ is *not* canceled by that of λ_r. $\Gamma_n^{(r)}$ actually does depend on μ. The general form of the *renormalization group equation* (RGE) exploits the μ-independence of $\Gamma_n^{(0)}$ to determine just how $\Gamma_n^{(r)}$ must depend on μ:

$$\mu \frac{\partial}{\partial \mu} \Gamma_n^{(0)} = 0 \tag{6.20}$$

$$\longrightarrow 0 = \mu \frac{\partial}{\partial \mu} \left(Z_\phi^{-n/2} \Gamma_n^{(r)}(\lambda_r, m_r, p_i, \mu) \right) \tag{6.21}$$

$$= Z_\phi^{-n/2} \left(\mu \frac{\partial}{\partial \mu} + \beta(\lambda) \frac{\partial}{\partial \lambda} + \gamma_m m \frac{\partial}{\partial m} - n\gamma(\lambda) \right) \Gamma_n^{(r)}(\lambda, m, p_i, \mu), \tag{6.22}$$

where the dimensionless functions are defined as

$$\beta(\lambda) = \mu \frac{\partial \lambda}{\partial \mu} \tag{6.23}$$

$$\gamma_m = \mu \frac{\partial}{\partial \mu} \ln m \tag{6.24}$$

$$\gamma(\lambda) = \mu \frac{\partial}{\partial \mu} \ln \sqrt{Z_\phi}. \tag{6.25}$$

The function $\gamma(\lambda)$ is called the *anomalous dimension*, for reasons that will become clear.

We want to use the renormalization group equation to deduce how $\Gamma_n^{(r)}$ changes if we rescale all the momenta:

$$p_i \to e^t p_i. \tag{6.26}$$

But since the procedure is a little complicated, I want to first illustrate what is going on using the 4-point function at one loop as an example. We had

$$\Gamma_4^{(r)}(p_i, \lambda_r, \mu) = \lambda_r \left(1 + \sum_{q^2 = s,t,u} \frac{\lambda_r}{32\pi^2} \ln \frac{q^2}{\mu^2} \right)$$

$$\to \Gamma_4^{(r)}(e^t p_i, \lambda_r, \mu) = \lambda_r \left(1 + \sum_{q^2 = s,t,u} \frac{\lambda_r}{32\pi^2} \ln \frac{e^t q^2}{\mu^2} \right)$$

$$= \Gamma_4^{(r)}(p_i, \lambda_r, e^{-t} \mu). \tag{6.27}$$

But we know that a change in μ can always be compensated by changing the value of λ_r[2]:

$$\Gamma_4^{(r)}(p_i, \lambda_r(\mu), \mu) = \Gamma_4^{(r)}(p_i, \lambda_r(e^t \mu), e^t \mu). \tag{6.28}$$

By simply renaming the slot for the explicit dependence on μ, this becomes

$$\Gamma_4^{(r)}(p_i, \lambda_r(\mu), e^{-t} \mu) = \Gamma_4^{(r)}(p_i, \lambda_r(e^t \mu), \mu). \tag{6.29}$$

Putting this together with (6.27), we get

---

[2] we can ignore the effect of w.f.r. here because we are only working at one loop in this example

$$\Gamma_4^{(r)}(e^t p_i, \lambda_r(\mu), \mu) = \Gamma_4^{(r)}(p_i, \lambda_r(e^t \mu), \mu). \qquad (6.30)$$

We see (again) that going to higher energy scales for the external particles is equivalent to changing the renormalized coupling constant, by rescaling μ. Recall that the μ dependence of λ_r was determined by integrating the beta function (6.23).

There were three things in the above example that made it special: (1) The 4-point function is dimensionless. (2) It did not depend on the mass. (3) There is no effect of wave function renormalization at one loop. Concerning (1), when the vertex has nonvanishing dimensions, then the first step, rescaling p_i by e^t, cannot be undone simply by absorbing it into μ. This is obvious with the 2-point function (inverse propagator) at tree level:

$$\Gamma_2 = p^2 \to e^{2t} p^2. \qquad (6.31)$$

In general, the dimension of $\Gamma_n^{(r)}$ is (mass)$^{4-n}$, i.e.,

$$\Gamma_n^{(r)} \sim |p_i|^{4-n} f\left(\lambda_r(\mu), \frac{p_i}{\mu}, \frac{m_r(\mu)}{\mu}, \mu\right) \qquad (6.32)$$

$$\to e^{(4-n)t} |p_i|^{4-n} f\left(\lambda_r(\mu), \frac{e^t p_i}{\mu}, \frac{m_r(\mu)}{\mu}, \mu\right) \qquad (6.33)$$

$$= e^{(4-n)t} |p_i|^{4-n} f\left(\lambda_r(\mu), \frac{p_i}{e^{-t}\mu}, \frac{e^{-t} m_r(\mu)}{e^{-t}\mu}, \mu\right), \qquad (6.34)$$

so that when we do the first step, we pick up a factor of $e^{(4-n)t}$ which wasn't there for the 4-point function. Concerning point (2), we see that the effect of e^t inside f can no longer be absorbed into a redefinition of the explicitly-appearing μ, since it can now be combined with both m and with p_i. And even if we could ignore the mass, point (3) implies that the function f is *not* invariant under a combined change of μ and λ_r, since we have to take w.f.r. into account as well. The final step is to use the μ-independence of the unrenormalized vertex to let $\mu \to e^t \mu$:

$$\Gamma_n^{(r)} \sim e^{(4-n)t} \underbrace{\left(\frac{Z_\phi(\lambda(e^t\mu), e^t\mu)}{Z_\phi(\lambda(\mu), \mu)}\right)^{-n/2}}_{e^{-n\int_0^t \gamma(\lambda_r(e^t\mu))dt}} |p_i|^{4-n} f\left(\lambda_r(e^t\mu), \frac{p_i}{\mu}, \frac{e^{-t} m_r(e^t\mu)}{\mu}, \mu\right).$$

$$\qquad (6.35)$$

Putting it all together, we get

$$\Gamma_n^{(r)}(e^t p_i, \lambda_r(\mu), m_r(\mu), \mu) = e^{(4-n)t - n\int_0^t \gamma(\lambda_r(e^t\mu))dt} \Gamma_n^{(r)}(p_i, \lambda_r(e^t\mu), e^{-t} m_r(e^t\mu), \mu).$$

$$\qquad (6.36)$$

This is the most general form of the solution to the RGE. Notice that I did not actually refer to the RGE to write it down; rather I used the μ-independence of the unrenormalized vertex. However, you can verify that (6.36) does satisfy the RGE. What was the point of differentiating $Z_\phi^{-n/2}\Gamma_n^{(r)}$ with respect to μ and then integrating again? Remember that by doing this for Γ_4, we were able to go from the simple one-loop result (6.10) for the effective coupling, to the RG-improved version (6.9) which summed up the leading logs. Integrating the RGE allows us to do likewise for the other proper vertices.

Now we can see the origin of the name *anomalous dimension*. If $\gamma(\lambda)$ was a constant, the overall rescaling factor in front of the vertex function would be

$$e^{(4-n-n\gamma)t}, \tag{6.37}$$

so having a nonzero value for γ is akin to changing the dimension of the vertex. It transforms as though its overall momentum dependence was of the form $|p_i|^{4-n-n\gamma}$ instead of $|p_i|^{4-n}$. This is the physical consequence of wave function renormalization.

Another point worth noting is that the presence of the mass generally makes it much more difficult to explicitly solve the RGE's in realistic problems, where we typically have several coupling constants whose RGE's are coupled to each other. If the external momenta are sufficiently large, then it is usually a good approximation to ignore the dependence on m.

Let's illustrate the RGE's for one more physical process, $2 \to 4$ scattering. How does $\Gamma_6(p_i)$ scale with changes in the energy scale $p_i \to e^t p_i$? Naively, we would expect the dependence

$$\Gamma_6(e^t p_i) \sim e^{-2t}\Gamma_6(p_i) \tag{6.38}$$

from the dimensionality of the process: the diagrams of figure 5.1 contain one propagator, which at high energies looks like $1/p^2$. But we know that the loop corrections to the vertices will alter this, in particular diagrams like Fig. 1.6b which introduce logarithmic dependence on the external momenta. If we ignore the mass dependence (assuming $p^2 \gg m^2$ in the virtual propagators) and the effect of w.f.r. (since this is higher order in λ), then the RGE tells us that

$$\Gamma_6(e^t p_i) \cong e^{-2t}\left(\frac{\lambda_r(e^t\mu)}{\lambda_r(\mu)}\right)^2 \Gamma_6(p_i), \tag{6.39}$$

since the tree diagram is of order λ^2. The significant thing is that the $\lambda_r(\mu)$ factors appearing here are the *renormalization group improved* ones (6.9), so indeed we have summed up the leading logs in getting this expression. If we further improve this by including the effect of w.f.r., we will see the anomalous dimension coming in, $e^{-2t} \to e^{-(2+\gamma)t}$ in the approximation where the anomalous dimension γ is relatively constant over the range of scales of interest. Finally, if we are working at scales where the masses are not negligible, we will have functions like $1/(p^2 - m_r^2(\mu))$ in the amplitude, which scales like $e^{-2t}/(p^2 - e^{-2t}m_r^2(e^t\mu))$.

Chapter 7
Other Regulators

The momentum space cutoff we have been using is the most intuitive form of regularization, but it is not the most efficient one. For a scalar field theory which does not have any gauge symmetries to preserve it is fine. But in a gauge theory, the momentum space cutoff is bad news. Under a gauge transformation, the vector potential and the electron field transform like

$$A_\mu(x) \rightarrow A_\mu(x) + \partial_\mu \Omega(x); \qquad \psi(x) \rightarrow e^{ie\Omega(x)}\psi(x). \qquad (7.1)$$

Suppose we have limited the quantum fluctuations of these fields to having (Euclidean) momenta with $|p| < \Lambda$. But if $\Omega(x)$ is rapidly varying in space, then the fields will have larger momenta, exceeding the cutoff, after the gauge transformation. In other words, the momentum cutoff does not respect the gauge symmetry. And breaking the gauge symmetry leads to all sorts of technical problems in defining the theory and insuring its consistency.

In the process of trying to define and prove the renormalizability of the SU(2) electroweak theory of the SM, 't Hooft and Veltman invented a very elegant alternative called *dimensional regularization* (DR). One regards the dimension of spacetime as a continuous parameter, under which physical quantities can be analytically continued. A loop diagram of the form

$$\int \frac{d^d p}{(2\pi)^d} \frac{1}{(p^2 + m^2)^\alpha} \qquad (7.2)$$

converges for $d < \alpha/2$. By evaluating loop integrals in the region of d where they converge, and then analytically continuing back to $d = 4$ (or close to 4), we are able to regularize the divergences. The expressions so obtained will have poles of the form $1/(d-4)$ which prevent us from taking $d \rightarrow 4$ until the divergences have been subtracted. So in this method, quantities which diverge as $\Lambda \rightarrow \infty$ using a cutoff display divergences in the form of $1/(d-4)^n$.

The original version of this chapter was revised: The errors in this chapter have been corrected. The correction to this chapter can be found at https://doi.org/10.1007/978-3-030-56168-0_16

© The Author(s), under exclusive license to Springer Nature Switzerland AG 2020, 37
corrected publication 2021
J. M. Cline, *Advanced Concepts in Quantum Field Theory*,
SpringerBriefs in Physics, https://doi.org/10.1007/978-3-030-56168-0_7

In scalar field theory, DR is very simple to implement; we just need to analytically continue the momentum space integral:

$$\frac{d^4 p}{(2\pi)^4} \to \frac{d\Omega_{d-1} p^{d-1} dp}{(2\pi)^d}. \tag{7.3}$$

There is a nice trick for deriving the volume Ω_{d-1} of the $d-1$-sphere. Evaluate the following integral (in Euclidean space) using either Cartesian or spherical coordinates:

$$\int \frac{d^d p}{(2\pi)^{d/2}} e^{-(p_1^2 + \cdots + p_d^2)/2} = \left(\int_{-\infty}^{\infty} \frac{dp}{(2\pi)^{1/2}} e^{-p^2/2} \right)^d = (1)^d \tag{7.4}$$

$$= \int \frac{d\Omega_{d-1} p^{d-1} dp}{(2\pi)^{d/2}} e^{-p^2/2} \tag{7.5}$$

$$= \frac{\Omega_{d-1}}{(2\pi)^{d/2}} 2^{d/2-1} \int_0^{\infty} du \, u^{d/2-1} e^{-u} \tag{7.6}$$

$$= \frac{1}{2} \frac{\Omega_{d-1}}{\pi^{d/2}} \Gamma(d/2). \tag{7.7}$$

Therefore

$$\Omega_{d-1} = \frac{2\pi^{d/2}}{\Gamma(d/2)}, \tag{7.8}$$

and we thus know how to define all the integrals that will arise.

In particular, the integral (7.2) evaluates to

$$\int \frac{d^d p}{(2\pi)^d} \frac{1}{(p^2 + m^2)^\alpha} = \frac{1}{2} \frac{\Omega_{d-1}}{(2\pi)^d} \int_0^{\infty} du \frac{u^{d/2-1}}{(u + m^2)^\alpha} \tag{7.9}$$

$$= \frac{1}{2} \frac{\Omega_{d-1}}{(2\pi)^d} (m^2)^{d/2-\alpha} \underbrace{\int_0^{\infty} dy \, y^{d/2-1} (1 + y)^{-\alpha}}_{\displaystyle \frac{\Gamma(d/2)\Gamma(\alpha - d/2)}{\Gamma(\alpha)}} \tag{7.10}$$

$$\frac{\Gamma(d/2)\Gamma(\alpha - d/2)}{\Gamma(\alpha)} \tag{7.11}$$

$$= \frac{\Gamma(\alpha - d/2)}{(4\pi)^{d/2}\Gamma(\alpha)} m^{d-2\alpha}. \tag{7.12}$$

In four dimensions, we know that the integral diverges for the common values of $\alpha = 1$ and 2. Let us write the number of dimensions as

$$d = 4 - 2\epsilon, \tag{7.13}$$

with the limit $\epsilon \to 0$ representing the removal of the cutoff (analogous to $\Lambda \to \infty$). The analytic properties of the gamma function tell us how the integral diverges. For positive integer values n,

$$\Gamma(-n + \epsilon) = \frac{(-1)^n}{n!} \left[\frac{1}{\epsilon} + \psi(n+1) + \frac{\epsilon}{2} \left(\frac{\pi^2}{3} + \psi^2(n+1) - \psi'(n+1) \right) + O(\epsilon^2) \right], \tag{7.14}$$

where ψ is the logarithmic derivative of the gamma function, $\psi(n+1) = 1 + 1/2 + \cdots + 1/n - \gamma$, $\psi'(n) = \pi^2/6 + \sum_{k=1}^n 1/k^2$, $\psi'(1) = \pi^2/6$ and $\gamma = \psi(1) = -0.5772\ldots$ is the Euler-Masheroni constant [5]. Normally one does not need to know all these finite parts; the $1/\epsilon$ pole is the most interesting part.

There is one more subtlety with DR: the action no longer has the right dimensions unless we adjust the coupling constant. The renormalized action now reads

$$S = \int d^d x \left(\frac{1}{2} (\partial \phi_0)^2 - \frac{1}{2} m_0^2 \phi_0^2 - \frac{\lambda_0}{4!} \phi_0^4 \right). \tag{7.15}$$

ϕ_0 has mass dimension of $d/2 - 1$ to make the kinetic term dimensionless; then the mass term is still fine, but λ_0 must now have mass dimension of $d - 2d + 4 = -d + 4 = 2\epsilon$. To make this explicit, we will keep λ_0 dimensionless, but introduce an arbitrary mass scale μ into the quartic coupling,

$$S \to \int d^d x \left(\frac{1}{2} (\partial \phi_0)^2 - \frac{1}{2} m_0^2 \phi_0^2 - \frac{\lambda_0}{4!} \mu^{2\epsilon} \phi_0^4 \right). \tag{7.16}$$

If we compute the 1-loop correction to the mass using DR, the result turns out to be

$$\begin{aligned} \delta m^2 &= \frac{\lambda m^2}{32\pi^2} \left(\frac{4\pi\mu^2}{m^2} \right)^{\epsilon} \Gamma(-1 + \epsilon) \\ &= \frac{\lambda m^2}{32\pi^2} \left(1 + \epsilon \ln \left(\frac{4\pi\mu^2}{m^2} \right) + \cdots \right) \left(-\frac{1}{\epsilon} + \psi(2) + \cdots \right) \\ &= \frac{\lambda m^2}{32\pi^2} \left(-\frac{1}{\epsilon} + \psi(2) - \ln \left(\frac{4\pi\mu^2}{m^2} \right) + O(\epsilon) \right). \end{aligned} \tag{7.17}$$

Similarly, the s-channel correction to the 4-point function can be computed, with the result

$$\delta\Gamma_4^{(a)}(q) = i\mu^{2\epsilon} \frac{\lambda_0^2}{32\pi^2} \left(\frac{1}{\epsilon} + \psi(1) + \int_0^1 dx \ln \left(\frac{4\pi\mu^2}{M^2(q)} \right) + O(\epsilon) \right), \tag{7.18}$$

where $M^2(q) = m^2 - q^2 x(1 - x)$ as in the momentum space cutoff computation, Eq. (5.10). Notice that we kept one factor of $\mu^{2\epsilon}$ in front, not expanding it in powers of ϵ, because we want Γ_4 to have the same dimension as the tree level diagram, $-i\lambda_0\mu^{2\epsilon}$. This factor will disappear at the end of any computation, after we have subtracted the divergences so that the cutoff can be removed, $\epsilon \to 0$.

In comparing the results in DR with those of the momentum space cutoff we can notice that the $1/\epsilon$ poles correspond exactly to the factors of $\ln \Lambda^2$ in the latter. In fact, at one loop the poles always appear in combination with $\ln 4\pi\mu^2$, and this quantity takes the place of $\ln \Lambda^2$. Recall that when we renormalized the parameters, the $\ln \Lambda$'s always got replaced by the log of the renormalization scale. In DR, the factors of $\ln \mu$ appear already in the loop diagrams, rather than being put into the counterterms. Nevertheless, the μ in DR plays exactly the same role as the renormalization scale we introduced previously.

To reiterate, the $1/\epsilon$ poles keep track of the log divergences. In an n-loop diagram, we will get leading divergences of the form $1/\epsilon^n$, and these will be associated with the leading logs. This is one reason DR is such an efficient regulator: the dominant divergences of any diagram allow one to immediately deduce the leading log contributions to running couplings. But what about the quadratic divergences, such as $\delta m^2 \sim \lambda\Lambda^2$? These are completely missing in DR. The process of analytic continuation has set them to zero. It is as though DR automatically provides the counterterms to exactly cancel the quadratic divergences, and only exhibits the logarithmic ones. This might seem like a way of avoiding the hierarchy problem for the Higgs boson mass, but our experience with the p-space cutoff tells us that the quadratic divergences must really be there, even if DR does not see them.

As for the finite parts of the diagrams, involving terms like $\psi(n)$, these are not very interesting, since the counterterms also have finite parts which could be chosen to cancel the $\psi(n)$'s. The different schemes for choosing finite parts for the counterterms are called *renormalization prescriptions*. There is one extremely simple prescription due to 't Hooft and Weinberg: take the counterterms to cancel the poles exactly, and nothing else. This is called *minimal subtraction* (MS). (There is also a variant called $\overline{\text{MS}}$ in which one subtracts all the factors of $\psi(1) + \ln 4\pi$ along with the $1/\epsilon$'s.) The detailed form of the RGE's is prescription-dependent, and they take a particularly simple form in the MS scheme.

Before leaving the subject of dimensional regularization, let's do one more computation, since it is a bit more subtle than the cutoff regulator. The beta function in dimensions other than 4 gets a new contribution of order ϵ. This is because the full bare coupling constant is $\lambda_{\text{bare}} = \lambda_0\mu^{2\epsilon}$, so even at tree level the renormalized coupling has μ dependence. The argument goes like this [7]. When we compute the counterterms to the coupling, we find

$$\lambda_{\text{bare}} = \mu^{2\epsilon}\lambda_r \left(1 + \frac{3\lambda}{32\pi^2\epsilon} + \text{finite} + \text{higher order} \right). \tag{7.19}$$

More generally we could write

$$\lambda_{\text{bare}} = \mu^{2\epsilon} \left(\lambda_r + \frac{a_1(\lambda)}{\epsilon} + \frac{a_2(\lambda)}{\epsilon^2} + \dots \right). \tag{7.20}$$

Now λ_{bare} is regarded as being independent of μ, which only comes in when we try to fix the value of the renormalized coupling. Therefore

$$\mu \frac{\partial}{\partial \mu} \lambda_{\text{bare}} = 0 = 2\epsilon \lambda_{\text{bare}} + \mu^{2\epsilon} \underbrace{\mu \frac{\partial}{\partial \mu} \lambda_r}_{\beta(\lambda_r)} \left(1 + \frac{a_1'(\lambda)}{\epsilon} + \frac{a_2'(\lambda)}{\epsilon^2} + \dots \right). \tag{7.21}$$

$$\beta(\lambda_r) \tag{7.22}$$

We can solve this for the beta function

$$\beta(\lambda_r, \epsilon) = -2\epsilon \left(\lambda_r + \frac{a_1(\lambda_r)}{\epsilon} - \lambda_r \frac{a_1'(\lambda_r)}{\epsilon} + \dots \right) = -2\epsilon \lambda_r + \frac{3\lambda_r^2}{16\pi^2} + \dots \tag{7.23}$$

Usually we will be interested in the limit $\epsilon \to 0$, but below we will show that this new term can have physical significance. Aside from the new term, this exercise illustrates how one computes the nonvanishing part of the beta function in DR. Interestingly, the terms of order $1/\epsilon^n$ in β which would blow up as $\epsilon \to 0$ must actually vanish due to relations between the a_i's. This harks back to our discussion of leading logs: the one-loop result fully determines the leading logs, and this is why the higher a_i's are determined by a_1.

Although DR is the most popular regulator on the market, there are others which are sometimes useful. I'll introduce a few of them briefly.

Pauli-Villars regularization is carried out by adding fictitious fields Φ_i with large masses M_i whose function is to cancel the divergences. Once this is done, the limit $M_i \to \infty$ can be taken, so that any low-energy effects of the fictitious fields vanish. In order to make the divergences due to the Φ_i's cancel those of the physical fields, one needs to make them anticommuting instead of commuting:

$$\{ \Phi_i(t, \mathbf{x}), \dot{\Phi}_j(t, \mathbf{y}) \} = i \delta_{i,j} \delta^{(3)}(\mathbf{x} - \mathbf{y}). \tag{7.24}$$

When we get to the discussion of fermions, we will learn that diagrams with a loop of anticommuting fields get an extra minus sign relative to those with commuting fields. Thus consider adding to the $\lambda \phi^4$ Lagrangian the terms

$$\frac{1}{2} \left((\partial \Phi)^2 - M^2 \Phi^2) \right) - \frac{\lambda}{4} \phi^2 \Phi^2 \tag{7.25}$$

for a single Pauli-Villars field Φ. It is not hard to verify that, with the normalization of the coupling as given, the combinatorics of the new diagrams are correct for cancelling the leading UV divergences in Γ_2 and Γ_4. Unfortunately, this only cures the quadratic divergences in Γ_2. For the log divergences, we are left with something proportional to $(M^2 - m^2) \log(\Lambda^2)$, which means we could not get away with just having one Pauli-Villars field. Since there are two kinds of divergences (quadratic

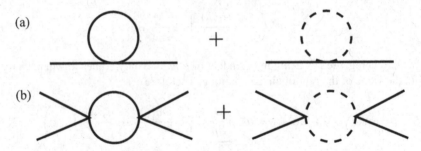

Fig. 7.1 Cancellation of divergences in **a** Γ_2 and **b** Γ_4 by Pauli-Villars field (dashed line)

and logarithmic) we need two PV fields, whose masses M_1 and M_2 are chosen in such a way as to eliminate both kinds (Fig. 7.1). In general one needs several fields to render all the loop integrals finite. Therefore this is rather messy compared to DR. However, it does have the virtue of preserving gauge invariance in gauge theories, and this is why it was commonly used before the invention of DR.

A variant on the sharp momentum space cutoff is to insert a smooth function $f(p, \Lambda)$ in the Euclidean space loop integrals, with the property that $\lim_{p\to\infty} f(p, \Lambda) = 0$ for fixed Λ and $\lim_{\Lambda\to\infty} f(p, \Lambda) = 1$ for fixed p. For example, $f = e^{-p^2/\Lambda^2}$ has this property.

Yet another way of regularizing loop diagrams is to give a parametric representation of the propagators (in Euclidean space),

$$\frac{1}{p^2 + m^2} = \lim_{\Lambda\to\infty} \int_{1/\Lambda^2}^{\infty} d\alpha \, e^{-\alpha(p^2+m^2)}. \tag{7.26}$$

For any $\alpha > 0$, the momentum integrals converge, and are easy to do because of the Gaussian factors. The hard work is postponed to performing the integrals over the α parameters. From these integrals we will get divergences of the same form as in the p-space cutoff method.

An extremely important form of regularization for nonperturbative computations is the *lattice cutoff*, where we imagine spacetime consists of discrete points on a hypercubic lattice with spacing a on each side. Every point in spacetime can be labeled by a set of integers n_μ, with $\mu = 0, 1, 2, 3$. The action (in Euclidean space) is discretized by writing

$$S = \sum_{n_\mu} a^4 \left(\frac{1}{2} \left[\sum_\nu \frac{1}{(2a)^2} (\phi_{n_\mu+e_\nu} - \phi_{n_\mu-e_\nu})^2 - m^2 \phi_{n_\mu}^2 \right] + V(\phi_{n_\mu}) \right), \tag{7.27}$$

where e_ν is the unit vector in the ν direction, and the fields are defined only on the points of the lattice. Going to momentum space, the kinetic term in the action becomes

$$S_{\text{kin}} = \frac{1}{2} \int \frac{d^4 p}{(2\pi)^4} \, \phi_p^* \left[\sum_\nu \left(\frac{e^{iap_\nu} - e^{-iap_\nu}}{2a} \right)^2 + m^2 \right] \phi_p, \qquad (7.28)$$

where the integration region is the hypercube defined by

$$-\frac{\pi}{a} \leq p_\nu \leq \frac{\pi}{a}. \qquad (7.29)$$

So $1/a$ plays the role of Λ. This method is not very useful for doing perturbative calculations since the propagator has a complicated form,

$$P(p) = \frac{1}{a^{-2} \sum_\nu \sin^2(ap_\nu) + m^2}, \qquad (7.30)$$

which makes loop integrals very difficult to do. Furthermore Lorentz invariance is broken by this method, and only restored in the continuum limit ($a \to 0$). Nevertheless, the lattice regulator is the only one which allows for a nonperturbative evaluation of the Feynman path integral. By discretizing spacetime, we have converted the functional integral into a regular finite-dimensional multiple integral, assuming we have also put the system into a finite box of size L^4. Then there are $(L/a)^4$ degrees of freedom, and the path integral can be evaluated on a computer, without perturbing in the coupling constant. This idea, applied to QCD, was one of the great contributions of K.G. Wilson.

A final method called zeta function regularization deserves mention, even though it is not useful for loop integrals. Rather, it comes into play when one wants to evaluate the Feynman path integral itself. Formally we can write

$$\int D\phi \, e^{\frac{i}{2} \int d^4 x ((\partial\phi)^2 - m^2\phi^2)} = \int D\phi \, e^{-\frac{i}{2} \int d^4 x \phi(\partial^2 + m^2)\phi} \qquad (7.31)$$

$$= \left(\det(\partial^2 + m^2) \right)^{-1/2}. \qquad (7.32)$$

This comes from the finite-dimensional formula

$$\int \frac{d^n x}{\sqrt{2\pi i}} \exp\left(\frac{i}{2} \sum_{a,b} x_a A_{a,b} x_b \right) = \int \frac{d^n y}{\sqrt{2\pi i}} \exp\left(\frac{i}{2} \sum_a \lambda_a y_a^2 \right) \qquad (7.33)$$

$$= \prod_{a=1}^n \sqrt{\frac{1}{\lambda_a}} = \sqrt{\frac{1}{\det A}}. \qquad (7.34)$$

To evaluate the determinant of an operator, we can use the famous identity that

$$\ln \det A = \operatorname{tr} \ln A = -\frac{d}{ds}\Big|_{s=0} \operatorname{tr} A^{-s} = \lim_{s\to 0}\frac{d}{ds}\underbrace{\sum_n \lambda_n^{-s}} \tag{7.35}$$

$$\equiv \lim_{s\to 0}\frac{d}{ds}\zeta_A(s), \tag{7.36}$$

where λ_n is the nth eigenvalue of the operator A. In QFT the sum is UV divergent because $A = p^2 - m^2$, but for sufficiently large values of s it converges. The nice thing about the zeta function $\zeta_A(s)$ associated with this operator is that its analytic continuation to $s = 0$ exists, so the limit is well defined.

The function $\zeta_A(s)$ associated with an operator A can be defined in terms of a "heat kernel," $G(x, y, \tau)$, which satisfies the diffusion-like equation

$$\frac{\partial}{\partial \tau} G(x, y, \tau) = A_x G(x, y, \tau), \tag{7.37}$$

where in the present application $A_x = \partial_x^2 + m^2$. The solution can be written in terms of the eigenvalues λ_n and the normalized eigenfunctions $f_n(x)$ as

$$G(x, y, \tau) = \sum_n e^{-\lambda_n \tau} f_n(x) f_n(y), \tag{7.38}$$

which has the property $G(x, y, 0) = \delta(x - y)$ by completeness of the eigenfunctions. The zeta function is defined as

$$\zeta_A(s) = \frac{1}{\Gamma(s)} \int_0^\infty d\tau\, \tau^{s-1} \int dx\, G(x, x, \tau) \tag{7.39}$$

$$= \frac{1}{\Gamma(s)} \sum_n \int_0^\infty d\tau\, \tau^{s-1} e^{-\lambda_n \tau} = \sum_n \lambda_n^{-s}. \tag{7.40}$$

Now the difficulty is in computing $G(x, x, \tau)$. For further details, see chapter III.5 of Ramond [5].

Chapter 8
Fixed Points and Asymptotic Freedom

In our derivation of the RG-improved coupling constant (6.9), we noted that $\lambda_r(\mu)$ grows with μ; in fact it blows up when μ reaches exponentially large values:

$$\lambda_r(\mu) \to \infty \quad \text{as} \quad \mu \to \mu_0 e^{\frac{16\pi^2}{3\lambda_r(\mu_0)}}, \tag{8.1}$$

which is called the Landau singularity—Landau observed that the same thing happens to the electron charge in QED. However, we cannot be sure that this behavior really happens unless we investigate the theory nonperturbatively. Since the coupling constant becomes large, perturbation theory is no longer reliable, and it is conceivable that $\lambda(\mu)$ could turn around at some scale and start decreasing again. Lattice studies of ϕ^4 theory have confirmed that the Landau singular behavior is in fact what happens. One finds that it is not possible to remove the cutoff in this kind of theory while keeping the effective coupling nonzero at low energies. This can be seen by summing the leading log contributions to the bare coupling:

$$\lambda_0 = \frac{\lambda_r}{1 - \frac{3\lambda_r}{16\pi^2} \ln \Lambda/\mu}. \tag{8.2}$$

Unless $\lambda_r = 0$, the coupling blows up at a finite value of Λ. If we insist on taking the limit $\Lambda \to \infty$, we must at the same time take $\lambda_r \to 0$. This means that the interactions disappear and we are left with a free field theory. This is what is meant by the *triviality* of ϕ^4 theory.

Triviality was considered a bad thing in the old days when it was felt that the ability to take the limit $\Lambda \to \infty$ should be a necessary requirement for a consistent field theory. That would be true if we demanded the theory to be a truly fundamental description. But if we are satisfied for it to be an effective theory, valid only up to some large but finite cutoff, there is no need to take $\Lambda \to \infty$. Thus ϕ^4 theory can still be physically meaningful despite its triviality.

© The Author(s), under exclusive license to Springer Nature Switzerland AG 2020
J. M. Cline, *Advanced Concepts in Quantum Field Theory*,
SpringerBriefs in Physics, https://doi.org/10.1007/978-3-030-56168-0_8

Here we see an example of a *fixed point* of the renormalization group: $\lambda_r(\mu) \to 0$ as $\mu \to 0$. We call it an *infrared stable* (IRS) fixed point since it occurs at low (IR) energy scales. This example happens to be a trivial fixed point. Much more interesting would be a nontrivial fixed point, where λ_r flows to some nonzero value as $\mu \to 0$. If this were to happen, it would lead to the startling conclusion that λ_r at low energies is quite insensitive to the value of the bare coupling we put into the theory. In this situation, the theory itself predicts the coupling, rather than us having to fix its value by adjusting a parameter.

There is one famous example of an IRS fixed point, due to K.G. Wilson and M.E. Fisher [8], which occurs in the ϕ^4 theory in $d < 4$ dimensions. We can find this using the β function (7.23), by noting that $\beta(\lambda_*)$ must vanish at an infrared fixed point λ_*,

$$\int_{\lambda_*} \frac{d\lambda}{\beta(\lambda)} = \int_{\mu=0} d\ln\mu = \infty, \qquad (8.3)$$

as well as at an ultraviolet one,

$$\int^{\lambda_*} \frac{d\lambda}{\beta(\lambda)} = \int^{\mu=\infty} d\ln\mu = \infty. \qquad (8.4)$$

Solving (7.23) for $\beta = 0$, we obtain the usual trivial fixed point at $\lambda_* = 0$, and in addition the nontrivial one

$$\lambda_* = \frac{32}{3}\pi^2\epsilon, \qquad (8.5)$$

which can be seen in Fig. 8.1. As advertised, this value is completely insensitive to the initial value of the coupling at very large renormalization scales, so it seems that we have indeed found a theory with the remarkable property that there are no free parameters determining the strength of the interaction at very low energies. We must be a little more careful however; remember that the full coupling constant is dimensionful in $4 - 2\epsilon$ dimensions:

$$\lambda_{*,\text{phys}} = \lambda_* \mu^{2\epsilon}. \qquad (8.6)$$

As we move deeper into the infrared, $\lambda_{*,\text{phys}}$ does get smaller even if the fixed point has been reached, but it does so in a way which is completely determined. Its absolute size is fixed by the dimensionless coupling λ_* and the energy scale μ at which we are measuring the coupling.

How do we know whether this is a UV or an IR fixed point? This depends on the sign of $\beta(\lambda)$ for λ near λ_*. If $\beta(\lambda) < 0$ for $\lambda < \lambda_*$ (this is true in the region $0 < \lambda < \lambda_*$ in Fig. 8.1a), then $\lambda(\mu)$ is a decreasing function of μ. λ *flows* away from the fixed point as μ increases, and toward it as μ decreases. Similarly If $\beta(\lambda) > 0$ for $\lambda > \lambda_*$ λ flows toward λ_* as μ decreases. Therefore the Wilson-Fisher fixed point is IR stable. I have shown the direction of flow of the coupling as μ *increases* in Fig. 8.1.

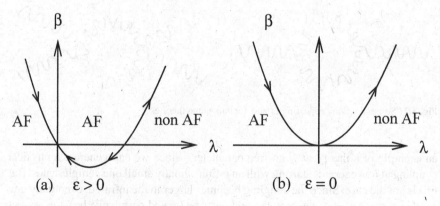

Fig. 8.1 $\beta(\lambda)$ for **(a)** $\epsilon > 0$ ($d < 4$) and **(b)** $\epsilon = 0$ ($d = 4$). Arrows show the direction of flow of λ in the UV. Asymptotically free regions are denoted by AF

For a nonvanishing physical value of ϵ like $\epsilon = 1$, the value of λ_* is quite large, and so one cannot necessarily trust the perturbative approximation that was used to derive it. However, it was shown by Wilson that the expansion in ϵ gives surprisingly accurate results even for such large values of ϵ, especially if one includes a few higher orders of corrections. This method has given useful results for statistical mechanics systems in three spatial dimensions at finite temperature. This works because in Euclidean space, the Lagrangian looks like a Hamiltonian, and the Feynman path integral becomes a partition function. After Wick rotating,

$$\int \mathcal{D}\phi \, e^{iS} \to \int \mathcal{D}\phi \, e^{-H/T}. \tag{8.7}$$

We can rewrite the Euclidean space Lagrangian as

$$\frac{1}{2}\left((\nabla\phi)^2 + m^2\phi^2\right) + \frac{1}{4!}\lambda\phi^4 = \frac{1}{T}\left(\frac{1}{2}\left((\nabla\tilde{\phi})^2 + m^2\tilde{\phi}^2\right) + \frac{1}{4!}\tilde{\lambda}\tilde{\phi}^4\right), \tag{8.8}$$

where $\phi = \tilde{\phi}/\sqrt{T}$ and $\lambda = \tilde{\lambda}T$.

We have seen an example of an IR fixed point, but what about UV fixed points? In Fig. 8.1a, the trivial fixed point is reached as $\mu \to \infty$, so this is a UV fixed point. The coupling λ flows toward zero in the ultraviolet. This is very different from the behavior in 4D when $\lambda > 0$, where we had $\lambda \to \infty$, which was the Landau singularity. It is an example of the very important phenomenon of *asymptotic freedom* (AF), so called because the theory becomes a free field theory for asymptotically large values of μ. Asymptotic freedom is a marvelous property from the point of view of perturbative calculability: the coupling gets weaker as one goes to higher energies, and perturbation theory becomes more accurate. Since such theories make sense up to arbitrarily large scales, they have the potential to be truly fundamental. Furthermore, it might seem that, just like in the case of an IR fixed point, this could provide

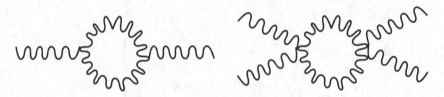

Fig. 8.2 Some diagrams appearing in nonAbelian gauge theories

an example of a theory with no free parameters, since we can generate a physical coupling at low energies starting with an infinitesimally small one at high scales. But this is not the case. Since the coupling becomes larger in the infrared, eventually one reaches a scale $\mu = \Lambda$ where it can no longer be treated pertubatively. Thus we get to trade what we thought was a dimensionless input parameter to the theory (λ) for a dimensionful one Λ, where in this context Λ is defined to be the energy scale at which $\lambda(\Lambda) \sim 1$. This exchange of a dimensionless for a dimensionful parameter is called *dimensional transmutation*.

In the case of $d = 4$, $\lambda < 0$, the ϕ^4 theory is AF, but since the potential is unbounded from below, this is not a physically interesting case. Until 1974 there were no convincing examples of asymptotic freedom in $d = 4$, but then it was discovered by 't Hooft, Politzer (working independently) and Gross and Wilczek that nonAbelian gauge theories, like the SU(3) of QCD, have a negative β function at one loop, and are therefore AF. Now that 't Hooft has received the Nobel Prize for proving the renormalizability of the electroweak theory, some expect that Politzer, Gross and Wilczek will soon be so honored for their discovery of AF. (At the time, Politzer was a graduate student at Harvard, and was the first to get the sign of the β function right after his senior competitors Gross and Wilczek at Princeton had initially made a sign error in their calculation.) The important difference between nonAbelian theories and U(1) theories like QED is the interaction of the gauge boson with itself, as in Fig. 8.2.

To make dimensional transmutation more explicit, the one-loop GR-improved coupling in QCD runs like

$$g^2(\mu) = \frac{g^2(\mu_0)}{1 + b_1 g^2(\mu_0) \ln(\mu^2/\mu_0^2)} \,, \tag{8.9}$$

and the beta function is

$$\beta(g) = -b_1 g^3 \,, \tag{8.10}$$

where for $n_f = 6$ flavors of quarks and $N = 3$ for SU(3) gauge theory, $b_1 = (11N/3 - 2n_f/3)/(8\pi^2) = 21/(24\pi^2)$. (The dependence on number of flavors comes from internal quark loops as in Fig. 8.3, which means that actually n_f also depends on μ, since a given quark will stop contributing when μ falls below its

Fig. 8.3 Quark loop
contribution to running of
QCD coupling

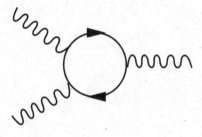

mass m_q.) Let us define a new scale Λ_{QCD} (not to be confused with the cutoff) such that

$$1 + 2b_1 g^2(\mu_0) \ln(\mu/\mu_0) \equiv 2b_1 g^2(\mu_0) \ln(\mu/\Lambda_{QCD}) \tag{8.11}$$

and

$$g^2(\mu) = \frac{1}{2b_1 \ln(\mu/\Lambda_{QCD})}, \tag{8.12}$$

which illustrates the claim that the renormalized coupling becomes independent of the value $g^2(\mu_0)$ at the reference scale. But the dimensionful parameter Λ_{QCD} still remains: it has become the free parameter in the theory. It has the interpretation of being the energy scale at which the coupling starts to become nonperturbatively large.

Another way of thinking about dimensional transmutation is that the theory has *spontaneously generated* a mass scale. This phenomenon is always associated with nonperturbative effects. Indeed, solving for Λ_{QCD} in terms of the coupling, we get

$$\Lambda_{QCD} = \mu e^{-\frac{1}{2b_1 g^2(\mu)}}, \tag{8.13}$$

which is independent of μ by construction. It is nonperturbative because $e^{-\frac{1}{2b_1 g^2(\mu)}}$ has no expansion in powers of $g^2(\mu)$. Λ_{QCD} cannot be predicted since it is the experimentally determined input parameter to the theory; its measured value is approximately 1.2 GeV, although the precise value depends on the renormalization scheme one chooses.

Chapter 9
The Quantum Effective Action

In problem 5, you showed that $iW[J] = \ln(Z[J])$ is the generator of connected Green's functions.[1] These of course are much more physically useful than the disconnected ones. But we have seen that the 1PI vertices are yet more important than the full set of connected diagrams. It is natural to wonder if there is generating functional which can give just the amputated 1PI diagrams. There is: it is called the *effective action*, $\Gamma[\phi_c]$. It is defined in terms of the proper vertices Γ_n by

$$\Gamma[\phi_c] = \sum_n \frac{1}{n!} \int \left(\prod_{j=1}^{n} d^4x_j \, \phi_c(x_j) \right) \Gamma_n(x_1, \ldots, x_n), \qquad (9.1)$$

where $\Gamma_n(x_1, \ldots, x_n)$ is the 1PI amputated n-point vertex, written in position space. Thus the ϕ_c factors mark the positions of the external lines in our usual way of computing diagrams.[2] The physical significance of $\Gamma[\phi]$ is that we can get *all* the diagrams of the theory, both 1PI and 1PR, by thinking of $\Gamma[\phi_c]$ as an effective action from which we construct tree diagrams only. For example, the complete 6-point amplitude has the form shown in Fig. 9.1, where each external leg is associated with a factor of ϕ_c, and there is no propagator for the external legs. The heavy dots represent the complete 4-point proper vertices and propagators summed to all orders in the loop expansion. One could imagine generating such diagrams from doing the path integral with $e^{i\Gamma[\phi_c]}$ but only including the tree diagrams.

There are different ways of computing $\Gamma[\phi_c]$. One which is especially nice for one-loop computations is the *background field* method [9]. We do it by splitting the field into two parts, a classical background field ϕ_c and a quantum fluctuation ϕ_q:

---

[1]Warning: I invert the meanings of W and Z relative to Ramond. I think this notation is more standard than his.

[2]Note: my earlier definition of Γ_n had the factor of i from $e^{i\Gamma}$ in the Γ_n's, whereas here I am defining Γ_n to be real.

The original version of this chapter was revised: The errors in this chapter have been corrected. The correction to this chapter can be found at https://doi.org/10.1007/978-3-030-56168-0_16

J. M. Cline, *Advanced Concepts in Quantum Field Theory*, SpringerBriefs in Physics, https://doi.org/10.1007/978-3-030-56168-0_9

Fig. 9.1 Generating all
diagrams from 1PI vertices
for the 6-point function

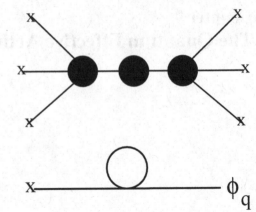

Fig. 9.2 Diagram which
mixes ϕ_c (the "x") with the
quantum flucutation ϕ_q

$$e^{i\Gamma[\phi_c]} = \int \mathcal{D}\phi_q\, e^{iS[\phi_c + \phi_q]}. \tag{9.2}$$

Let us now expand S in powers of ϕ_q. At zeroth order, we get $S[\phi_c]$, which can be brought outside the functional integral. At first order, we get

$$\phi_q \left(-\partial^2 - m^2 - \frac{\lambda}{3!}\phi_c^2 \right) \phi_c \equiv 0. \tag{9.3}$$

To perform the functional integral, it is convenient to make this term vanish so there is no tadpole for ϕ_q. It does vanish provided that ϕ_c satisfies its classical equation of motion, hence the identification of ϕ_c as the classical field. Now when we rewrite the integrand, we get

$$e^{i\Gamma[\phi_c]} = e^{iS_0[\phi_c]} \int \mathcal{D}\phi_q \exp\left(iS_0[\phi_q] - i\int d^4x[V(\phi_c + \phi_q) - \phi_q V'(\phi_c)]\right), \tag{9.4}$$

where S_0 is the free part of the action. The subtracted part is removed due to the interaction term in the equation of motion for ϕ_c. This is the key to getting rid of the 1PR diagrams: they are diagrams constructed from vertices that have only a single internal line. Since all the internal lines are due to ϕ_q, we eliminate such diagrams by removing the term linear in ϕ_q from the interaction.

How this works at tree level is clear, but in loop diagrams it is possible to induce mixing between ϕ_q and ϕ_c which feeds into 1PR diagrams. In ϕ^4 theory such mixing starts at one loop, as shown in Fig. 9.2, and it will give rise to corrections like Fig. 9.3 which are no longer 1PI.

The reason is that once we introduce loops, the classical equation will get quantum corrections; thus (9.3) is no longer quite right. At one loop we get the correction

Fig. 9.3 1PR diagram
induced by the mixing
shown in Fig. 9.3

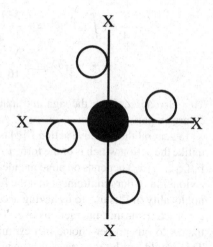

$$\phi_q \left(-\partial^2 - m^2 - \frac{\lambda}{3!}\phi_c^2 - \frac{\lambda}{3!}\langle\phi_q^2\rangle \right) \phi_c = 0 \,. \tag{9.5}$$

Adding this correction will take care of the unwanted mixing terms. Let's now show how things work at lowest nontrivial order using the background field method.

I often find this to be a nice way of keeping track of the signs of the quantum corrections to mass and coupling. For the mass correction in ϕ^4 theory, we can write

$$i\Gamma \cong \int \mathcal{D}\phi\, e^{iS_0} \int d^4x \left(-i\frac{\lambda}{4!}\frac{4!}{2!2!}\phi_c^2\phi_q^2 \right), \tag{9.6}$$

$$\longrightarrow -\frac{i}{2}\delta m^2 \phi_c^2 = -\frac{i\lambda}{4}\phi_c^2 G_2(x,x) \tag{9.7}$$

$$\longrightarrow \delta m^2 = \frac{\lambda}{2} \int \frac{d^4p}{(2\pi)^4} \frac{i}{p^2 - m^2 + i\varepsilon}, \tag{9.8}$$

which agrees with our previous calculation (2.4). This way of determining the sign is more direct than our previous reasoning. Similarly we can obtain $\delta\lambda$:

$$i\Gamma \cong \int \mathcal{D}\phi\, e^{iS_0} \frac{1}{2}\left(\int d^4x \left(-i\frac{\lambda}{4!}\frac{4!}{2!2!}\phi_c^2\phi_q^2 \right) \right)^2 \tag{9.9}$$

$$\longrightarrow -\frac{i}{4!}\delta\lambda \int d^4x\, \phi_c^4 = -\frac{\lambda^2}{16} \int d^4x \int d^4y\, \phi_c^2(x)\phi_c^2(y) G_2(x-y)^2 \,. \tag{9.10}$$

We can Taylor-expand $\phi_c^2(y) = \phi_c^2(x) + (y - x) \cdot \partial\phi_c^2(x) + \cdots$. The higher order terms represent the momentum dependent parts of the 4-point function, which we could keep track of if we wanted to, but if we just want to confirm the p-independent part of the coupling, we can neglect these. Then

$$-\frac{i}{4!}\delta\lambda \int d^4x\,\phi_c^4 = -\frac{\lambda^2}{16}\int d^4x\,\phi_c^4(x)\int d^4y\,G_2(x-y)^2 \qquad (9.11)$$

$$\longrightarrow -\frac{i}{4!}\delta\lambda = -\frac{\lambda^2}{16}\int \frac{d^4p}{(2\pi)^4}\left(\frac{i}{p^2-m^2+i\varepsilon}\right)^2. \qquad (9.12)$$

This also agrees with the sign and magnitude of our previous computation of the divergent part of $\delta\lambda$.

In general the effective action $\Gamma[\phi]$ is a *nonlocal* functional of the classical fields: unlike the action which is of the form $\int d^4x\,\mathcal{L}(x)$, which has exclusively local terms, $\Gamma_n(x_1,\dots,x_n)$ depends on noncoincident positions of the fields. This is of course not a violation of our requirement that the fundamental interactions should be local. The nonlocality in Γ is due to us having accounted for the effects of propagation: virtual ϕ particles transmit interactions over a distance, in a way which respects causality, thanks to our proper choice of Feynman boundary conditions in the propagators. This would not be the case for a generic function of the x_i's we might randomly write down. That is an important distinction between the underlying theory and the effective theory.

Since we can always reexpress nonlocalities in terms of higher derivatives using Taylor expansions, another way of writing Γ which makes it look more like a local functional is

$$\Gamma[\phi_c] = \int d^4x\left[-V_0(\phi_c)+V_2(\phi_c)(\partial\phi_c)^2+V_4(\phi_c)(\partial\phi_c)^4+\cdots\right]. \qquad (9.13)$$

If we now specialize to constant fields, the derivative terms drop out and we are left with only V_0, which is called the *effective potential*. Let's give one very famous example [10] due to Coleman and Weinberg. They noticed that it is possible to sum up all the one-loop terms to all orders in ϕ_c because of the fact that these terms just amount to a field-dependent shift in the mass. This is illustrated in Fig. 9.4, where the x's represent ϕ_c's (with no internal propagator). The diagram $\bigcirc$ is the value of $\ln(Z[0])/i$, that is, the log of free path integral with no insertions of the external field:

$$\begin{aligned}
\bigcirc &= -i\,\ln\left(\det^{-1/2}(-\partial^2-m^2)\right)\\
&= \frac{i}{2}\,\mathrm{tr}\,\ln(-\partial^2-m^2)\\
&= \frac{i}{2}\int\frac{d^4p}{(2\pi)^4}\langle p|\ln(p^2-m^2)|p\rangle\\
&= \frac{i}{2}\langle p|p\rangle\int\frac{d^4p}{(2\pi)^4}\ln(p^2-m^2+i\varepsilon)\\
&= \frac{i}{2}\int d^4x\int\frac{d^4p}{(2\pi)^4}\ln(p^2-m^2+i\varepsilon).
\end{aligned}$$

$$(9.14)$$

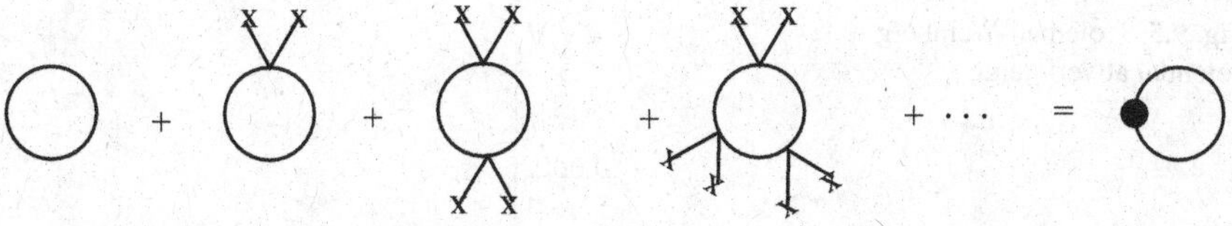

Fig. 9.4 Resummation for Coleman-Weinberg potential

Here we have used that fact that $\langle p|q \rangle = \int d^4 x e^{i(p-q)\cdot x}$ to get the factor of the volume of spacetime. Now the claim is that when we add up the series in Fig. 9.4, we get precisely

$$\begin{aligned}
\bullet &= -\int d^4 x \, V_{\rm cw}(\phi_c) = \frac{i}{2} \int d^4 x \int \frac{d^4 p}{(2\pi)^4} \ln(p^2 - m^2 - \lambda\phi_c^2/2 + i\varepsilon) \\
&= -\frac{1}{2} \int d^4 x \int \frac{d^4 p_E}{(2\pi)^4} \ln(p_E^2 + m^2 + \lambda\phi_c^2/2) + {\rm const}.
\end{aligned} \tag{9.15}$$

To see that it is true, one just has to expand (9.15) as a power series in λ and compare each term to the corresponding Feynman diagram. The infinite imaginary constant from $\ln(-1)$ has no physical significance since we could have avoided it by normalizing our path integral measure appropriately.

To evaluate the Coleman-Weinberg potential, it is convenient to first differentiate it with respect to m^2; we can then use our favorite regulator to evaluate it. With the momentum space cutoff, and defining $M^2 = m^2 + \lambda\phi_c^2/2$,

$$\begin{aligned}
\frac{\partial V_{\rm cw}}{\partial M^2} &= \frac{1}{2} \int \frac{d^4 p_E}{(2\pi)^4} \frac{1}{p_E^2 + m^2 + \lambda\phi_c^2/2} \\
&= \frac{1}{32\pi^2} \int_{M^2}^{\Lambda^2 + M^2} du \frac{u - M^2}{u} \\
&= \frac{1}{32\pi^2} \left(\Lambda^2 - M^2 \ln(1 + \Lambda^2/M^2) \right).
\end{aligned} \tag{9.16}$$

Ignoring terms which vanish as $\Lambda \to \infty$ and an irrelevant constant, the integrated potential is

$$V_{\rm cw} = \frac{1}{64\pi^2} \left(2\Lambda^2 M^2 - M^4 \ln \left(\frac{\Lambda^2}{M^2} \right) - \frac{1}{2} M^4 \right). \tag{9.17}$$

You can easily see that the only divergent terms if we reexpand this in powers of ϕ_c are the ϕ_c^2 and ϕ_c^4 terms, and moreover the log divergent shift in the ϕ_c^4 term agrees with our previous calculations. The quadratically divergent part can be completely removed by renormalizing the mass (and the cosmological constant, but we shall ignore the latter), so the M^4 term is the most interesting one. After renormalization, and adding the one-loop result to the tree-level potential, we obtain

Fig. 9.5 Coleman-Weinberg
potential at very large μ

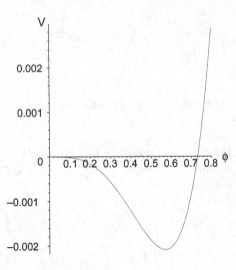

$$V(\phi_c) = \frac{1}{2}m^2\phi^2 + \frac{1}{4!}\lambda\phi^4 + \frac{1}{64\pi^2}\left(m^2 + \frac{1}{2}\lambda\phi^2\right)^2 \ln\left(\frac{m^2 + \frac{1}{2}\lambda\phi^2}{\mu^2}\right). \quad (9.18)$$

I have defined μ here in such a way as to absorb the M^4 which is not logarithmic into the log. We can see the μ-dependence of the effective coupling constant by taking

$$\lambda_{\text{eff}} = \left.\frac{\partial^4 V}{\partial\phi_c^4}\right|_{\phi_c=0} = \lambda + \frac{3\lambda^2}{32\pi^2}\left(\ln\left(\frac{m^2}{\mu^2}\right) + \frac{3}{2}\right). \quad (9.19)$$

Notice that since we are computing the effective potential, we have set all external momenta to zero. Therefore, roughly speaking, the coupling is evaluated at the lowest possible energy scale in the theory, m^2. This result thus has the same basic form as our energy-dependent effective coupling (6.6).

A curiosity about the Coleman-Weinberg potential can be observed for sufficiently large values of μ: the minimum of the potential is no longer at $\phi = 0$, but at some nonzero value, as illustrated in Fig. 9.5. If this behavior occurred, it would be an example of *spontaneous symmetry breaking*. Although the tree-level potential has the symmetry $\phi \to -\phi$, the quantum theory gives rise to a nonzero *vacuum expectation value* (VEV), $\langle\phi\rangle$, which breaks the symmetry. In the present case, we call it *radiative symmetry breaking* since it is induced by radiative corrections—another term for loop corrections. Spontaneous symmetry breaking is an important feature of the standard model; the Higgs field gets a VEV which breaks the SU(2)×U(1) gauge group to the U(1) of electromagnetism, and gives masses to the fermions.

It would be interesting if a theory gave rise to spontaneous symmetry breaking without having to put in a negative value for m^2 by hand, which is the usual procedure. Unfortunately, the Coleman-Weinberg potential cannot be relied upon to do this since the value of μ required is nonperturbatively large. For simplicity we consider the massless case:

$$\mu^2 \cong \frac{\lambda}{2} \langle\phi\rangle^2 \exp\left(\frac{32\pi^2}{\lambda}\right). \tag{9.20}$$

For such large values of μ, the loop correction is so big that we have already reached the Landau singularity. This can be seen by RG-improving the one-loop result to get the log into the denominator. Thus the calculation cannot be trusted in this interesting regime. However, it is possible to get radiative symmetry breaking in more complicated models, in a way that does not require going beyond the regime of validity of perturbation theory.

Now I would like to discuss a more formal definition of the effective action, in which Γ is related to $W[J]$ through a functional Legendre transformation. In this approach we define the classical field ϕ_c by

$$\phi_c[J] = \frac{\delta W}{\delta J}, \tag{9.21}$$

where we do not set $J = 0$ (until the end of the calculation). ϕ_c is the expectation value of ϕ when there is a source term $+\int d^4x\, \phi J$ in the action; thus it satisfies the classical equation

$$\left(\partial^2 + m^2 + \frac{\lambda}{2}\phi_c^2\right)\phi_c(x) = J(x) \tag{9.22}$$

plus the quantum corrections like we derived in Eq. (9.5). Hence

$$\begin{aligned}
\phi_c &= \left(-\partial^2 - m^2 - \frac{\lambda}{2}\phi_c^2\right)^{-1}(-J) \\
&= \left[(-\partial^2 - m^2)\left(1 - (-\partial^2 - m^2)^{-1}\frac{\lambda}{2}\phi_c^2\right)\right]^{-1}(-J) \\
&= -\left[G_2 \cdot J + G_2 \cdot \frac{\lambda}{2}\phi_c^2 \cdot G_2 \cdot J + G_2 \cdot \frac{\lambda}{2}\phi_c^2 \cdot G_2 \cdot \frac{\lambda}{2}\phi_c^2 \cdot G_2 \cdot J + \cdots\right],
\end{aligned} \tag{9.23}$$

where I am using the notation[3]

$$G_2 \cdot J = \int d^4y\, G_2(x-y)J(y)$$

$$G_2 \cdot \frac{\lambda}{2}\phi_c^2 \cdot G_2 \cdot J = \int d^4y \int d^4z\, G_2(x-y)\frac{\lambda}{2}\phi_c^2(y)G_2(y-z)J(z) \quad etc. \tag{9.24}$$

This could be evaluated perturbatively by taking $\phi_0 = G_2 \cdot J$, substituting this into (9.23) to obtain $\phi_1 = \phi_0 + G_2 \cdot \frac{\lambda}{2}\phi_0^2 \cdot G_2 \cdot J$, and continuing this process to the

---

[3]Note: I am henceforth defining G_2 to be the inverse of $(-\partial^2 - m^2)$, which in momentum space looks like $1/(p^2 - m^2)$, not $i/(p^2 - m^2)$. Furthermore I am redefining the source term in the Lagrangian to be $+J\phi$ in order to match the convention which is more standard in the derivation of the effective action below.

desired order in λ. However we won't need an explicit solution in order to carry out the proof of the following claim: the effective action is related to $W[J]$ by

$$\Gamma[\phi_c] = W[J] - J \cdot \phi_c. \tag{9.25}$$

To prove this, we will start functionally differentiating $\Gamma[\phi]$ (let's drop the subscript c and replace it with $\phi_c(x) = \phi_x$ when necessary) with respect to ϕ:

$$\frac{\delta \Gamma}{\delta \phi} = \frac{\delta W}{\delta J} \cdot \frac{\delta J}{\delta \phi} - \frac{\delta J}{\delta \phi} \cdot \phi - J = -J. \tag{9.26}$$

The latter equality follows from the fact that $\frac{\delta W}{\delta J} = \phi_c$. In passing, we note that $\frac{\delta \Gamma}{\delta J} = 0$ when $J = 0$; similarly $\frac{\delta W}{\delta \phi} = 0$ when $\phi = 0$.

Next let's differentiate again:

$$\frac{\delta^2 \Gamma}{\delta \phi_x \delta \phi_y} = -\frac{\delta J_x}{\delta \phi_y} = -\left[\frac{\delta \phi_x}{\delta J_y}\right]^{-1} = -\left[\frac{\delta^2 W}{\delta J_x \delta J_y}\right]^{-1}. \tag{9.27}$$

If we recall problem 5,[4]

$$\frac{\delta^2 W}{\delta J_x \delta J_y} = -i \left\langle (i\phi_x)(i\phi_y) \right\rangle_{\text{con}}$$
$$= G_2^{\text{con}}(x - y). \tag{9.29}$$

Here $G_2^{con}(x - y)$ is the connected part of the full propagator, to all orders in perturbation theory. Because Γ_2 is the full propagator with both legs multiplied by inverse propagators, we see that

$$\frac{\delta^2 \Gamma}{\delta \phi_x \delta \phi_y} = \Gamma_2[\phi], \tag{9.30}$$

as promised. Notice that the inverse propagator is automatically 1PI, in analogy to the fact that $(1 + x + x^2 + \cdots)^{-1} = 1 - x$.

We can continue this procedure to find higher derivatives of $\Gamma[\phi]$ and inductively prove the desired relation. Things get a little more complicated as we go to higher point functions. To obtain Γ_3 (which does not exist in ϕ^4 theory, but it does in ϕ^3 theory), it is useful to differentiate relation (9.21) with respect to ϕ [11]:

---

[4] Note the important generalization that

$$\frac{\delta^n W}{\delta J_1 \cdots \delta J_n} = -i^{n+1} \langle T(\phi_1 \cdots \phi_n) \rangle_{\text{con}} \tag{9.28}$$

which follows from the fact that $W = -i \ln Z$, so $\frac{\delta^n W}{\delta J_1 \cdots \delta J_n} = -i Z^{-1} \frac{\delta^n Z}{\delta J_1 \cdots \delta J_n} +$ (terms that remove the disconnected parts) $= -i \langle T(i\phi_1 \cdots i\phi_n) \rangle_{\text{con}}$

$$\frac{\delta^2 W}{\delta J_x \delta J_y} \cdot \frac{\delta J_y}{\delta \phi_z} = \delta_{x,z} \,. \qquad (9.31)$$

Using (9.27) we can substitute for $\frac{\delta J_y}{\delta \phi_z}$ to write this as

$$\frac{\delta^2 W}{\delta J_x \delta J_y} \cdot \frac{\delta^2 \Gamma}{\delta \phi_y \delta \phi_z} = -\delta_{x,z} \qquad (9.32)$$

and take a derivative with respect to J:

$$\frac{\delta^3 W}{\delta J_w \delta J_x \delta J_y} \cdot \frac{\delta^2 \Gamma}{\delta \phi_y \delta \phi_z} + \frac{\delta^2 W}{\delta J_x \delta J_y} \cdot \frac{\delta \phi_y}{\delta J_u} \cdot \frac{\delta^3 \Gamma}{\delta \phi_u \delta \phi_y \delta \phi_z} = 0 \,. \qquad (9.33)$$

Now using (9.31) and (9.32) this can be solved for $\frac{\delta^3 \Gamma}{\delta \phi_u \delta \phi_y \delta \phi_z}$:

$$\frac{\delta^3 \Gamma}{\delta \phi_u \delta \phi_v \delta \phi_w} = -\int d^4 x \int d^4 y \int d^4 z \, \frac{\delta^3 W}{\delta J_x \delta J_y \delta J_z} \frac{\delta^2 \Gamma}{\delta \phi_x \delta \phi_u} \frac{\delta^2 \Gamma}{\delta \phi_y \delta \phi_v} \frac{\delta^2 \Gamma}{\delta \phi_z \delta \phi_w} \,. \qquad (9.34)$$

To see why the $-$ sign should be there, note that eq. (9.28) implies that $\frac{\delta^3 W}{\delta J_x \delta J_y \delta J_z} = -\langle T(\phi_x \phi_y \phi_z) \rangle_{\mathrm{con}}$. For the $\frac{1}{3!} \mu \phi^3$ interaction, we would thus get (in momentum space)

$$\frac{\delta^3 W}{\delta J_1 \delta J_2 \delta J_3} \frac{\delta^2 \Gamma}{\delta \phi_1 \delta \phi_1} \frac{\delta^2 \Gamma}{\delta \phi_2 \delta \phi_2} \frac{\delta^2 \Gamma}{\delta \phi_3 \delta \phi_3} = -(-i\mu) \frac{i}{p_1^2 - m^2} \frac{i}{p_2^2 - m^2} \frac{i}{p_3^2 - m^2}$$
$$\times (p_1^2 - m^2)(p_2^2 - m^2)(p_3^2 - m^2)$$
$$= \mu \,, \qquad (9.35)$$

whereas we know that $\frac{\delta^3 \Gamma}{\delta \phi_1 \delta \phi_1 \delta \phi_3} = -\mu$ (since the action is the kinetic term minus the potential). Therefore the right hand side of (9.34) is indeed the connected 3-point function with external propagators removed by the $\frac{\delta^2 \Gamma}{\delta \phi \delta \phi}$ factors, which is the same as the proper vertex Γ_3. I will leave as an exercise for you to show how things work for the 4-point function.

In the previous manipulations we have been referring to the effective action at all orders in the loop expansion. However, it is also possible to truncate it, and work with the 1-loop or 2-loop effective action. To reiterate: the complete result for any amplitude at this order will be given by the tree diagrams constructed from Γ.

I will quote one other interesting result from Ramond [5] without deriving it. It can be shown that the equation satisfied by the classical field is

$$(\partial^2 + m^2)\phi_x - J_x = -\frac{1}{Z[J]} \frac{\partial V}{\partial \phi} \left(-i \frac{\delta}{\delta J_x} \right) Z[J] \qquad (9.36)$$

where the argument of V' instead of being ϕ_x is the operator $-i\frac{\delta}{\delta J_x}$. This is a general result; specializing to ϕ^4 theory, we get

$$(\partial^2 + m^2)\phi_x - J_x = -\frac{\lambda}{3!}\phi_x^3 + i\frac{\lambda}{4}\frac{\delta\phi_x^2}{\delta J_x} + \frac{\lambda}{3!}\frac{\delta^2\phi_x}{\delta J_x \delta J_x} \qquad (9.37)$$

By restoring factors of $\hbar$, one can show that the three terms on the r.h.s. of (9.37) are of order $\hbar^0$, $\hbar^1$ and $\hbar^2$ respectively. Therefore the last term arises at two loops. It can be seen that the quantum correction we derived in (9.5) does correspond to the one-loop term in (9.37).

To conclude this chapter, I would like to mention another kind of effective action which is due to Wilson [8]. Suppose we have computed the renormalized Lagrangian $\mathcal{L}$ for some particular value of the cutoff Λ. We can define the effective value of $\mathcal{L}$ at a lower scale, Λ', by splitting the field in Fourier space into two pieces:

$$\tilde{\phi}(p) = \begin{cases} \tilde{\phi}_{IR}, & 0 < |p| < \Lambda' \\ \tilde{\phi}_{UV}, & \Lambda' < |p| < \Lambda \end{cases}. \qquad (9.38)$$

The corresponding fields in position space are obtained by doing the inverse Fourier transform. It is easiest to define what we mean by $|p|$ if we are working in Euclidean space. The Wilsonian effective action, at the scale Λ', is given by integrating over the UV degrees of freedom only:

$$e^{-S'[\phi_{IR}; \Lambda']} = \int \mathcal{D}\phi_{UV}\, e^{-S[\phi_{IR}+\phi_{UV}; \Lambda]}. \qquad (9.39)$$

In practice, this is accomplished by integrating all Feynman diagrams over the region $\Lambda' < |p| < \Lambda$ instead of $0 < |p| < \Lambda$. Notice that this way of splitting the field into UV and IR parts respects our desire to have no mixing between ϕ_{IR} and ϕ_{UV} in the kinetic term, since it is diagonal in the momentum basis. Thus ϕ_{IR} is similar to ϕ_c and ϕ_{UV} is like ϕ_q in the above discussion.

This process of *integrating out* the dynamical degrees of freedom above a certain scale is Wilson's way of defining a RG transformation. Suppose the original Euclidean action had been

$$S[\phi; \Lambda] = \frac{1}{2}\left((\partial\phi)^2 + m^2(\Lambda)\phi^2\right) + \frac{\lambda(\Lambda)}{4!}\phi^4; \qquad (9.40)$$

then we expect the effective action to have the form

$$S'[\phi; \Lambda'] = \frac{1}{2}\left(\left(1 + \delta Z\left(\frac{\phi}{\Lambda'}\right)\right)(\partial\phi)^2 + \hat{m}^2(\Lambda')\phi^2\right) + O\left(\frac{(\partial\phi)^4}{\Lambda'^4}\right) + \frac{\hat{\lambda}(\Lambda')}{4!}\phi^4 + O\left(\frac{\phi^6}{\Lambda^2}\right). \qquad (9.41)$$

To make this look more like the original Lagrangian, we should renormalize $\phi = \phi'/\sqrt{1+\delta Z}$:

$$S'[\phi'; \Lambda'] = \frac{1}{2}\left((\partial\phi')^2 + m^2(\Lambda')\phi'^2\right) + O\left(\frac{(\partial\phi')^4}{\Lambda'^4}\right) + \frac{\lambda(\Lambda')}{4!}\phi^4 + O\left(\frac{\phi'^6}{\Lambda'^2}\right).$$
(9.42)

In this approach, the cutoff is now playing the role of the renormalization scale. The parameters like λ' appearing in S' are telling us what physics looks like at the scale Λ'. It is not hard to see why when considering loop diagrams like Fig. 3.1: if all the external legs have large momenta, then it is impossible to get contributions from the loops where the internal lines have small momenta. There is only a set of measure zero which conserves momentum at the vertices while restricting $|p| < \Lambda'$. So we can use just the tree diagrams generated by $S'[\phi'; \Lambda']$ to compute physical amplitudes at the scale of the cutoff. This is similar in spirit though different in detail to the way the quantum effective action works. In particular, the Wilsonian effective action *does* contain the effects of 1PR diagrams, whereas these have to be constructed from the proper vertices of the quantum effective action.

Wilson does one further step in his way of defining $S'[\phi'; \Lambda']$. Notice that even in the case of free field theory, S' does not keep exactly the same form as S in the way we have so far defined it, since Λ appears in the upper limit of momentum integrals:

$$S_0[\phi; \Lambda] = \frac{1}{2}\int_0^\Lambda \frac{d^d p}{(2\pi)^4}\tilde{\phi}_{-p}(p^2 + m^2\phi^2)\tilde{\phi}_p.$$
(9.43)

To make S_0 appear really independent of the cutoff, we can define a dimensionless momentum $q = p/\Lambda$:

$$S_0[\phi; \Lambda] = \Lambda^d\frac{1}{2}\int_0^1 \frac{d^d q}{(2\pi)^4}\tilde{\phi}_{-q}(\Lambda^2 q^2 + m^2)\tilde{\phi}_q$$
(9.44)

and rescale the fields by $\phi \to \zeta_0\phi$ where $\zeta_0 = \Lambda^{-d/2-1}$:

$$S_0[\phi; \Lambda] \to \frac{1}{2}\int_0^1 \frac{d^d q}{(2\pi)^4}\tilde{\phi}_{-q}(q^2 + (m^2/\Lambda^2))\tilde{\phi}_q.$$
(9.45)

This is equivalent to taking the unit of energy to be Λ at any stage in the renormalization procedure. When we integrate out momenta in the range $\Lambda' < |p| < \Lambda$, we will pick up an additional factor of $(\Lambda'/\Lambda)^{-d/2-1}$ in the step where we renormalize the fields. One notices that the effective mass m^2/Λ^2 gets continously bigger as one integrates out more fluctuations. We call ϕ^2 a *relevant operator* because its coefficient gets bigger as we go deeper into the infrared. On the other hand, consider the coefficient of an operator of the form ϕ^n:

Fig. 9.6 Contribution to the
6-point function in
Wilsonian effective action

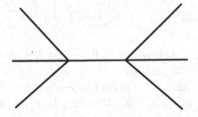

$$S_{\text{int}} = c_n \prod_{i=1}^{n} \left(\int_0^{\Lambda} \frac{d^d p}{(2\pi)^4} \tilde{\phi}_{p_i} \right) (2\pi)^d \delta^d \left(\sum_i^n p_i \right) \tag{9.46}$$

scales like $\Lambda^{(n-1)d} \zeta^n \sim \Lambda^{-nd/2+d+n}$. For $d = 4$ this is Λ^{-d+n}, which we recognize to be simply the dimension of the operator. An operator which does not scale with Λ, such as $\lambda\phi^4$ in 4D, is called *marginal*. Its effects neither become very strong nor very weak in the infrared. Of course this naive argument does not take into account the running due to quantum effects. If the coupling runs logarithmically to larger values in the IR, as is the case in QCD, then it is called *marginally relevant*, and if it runs to smaller values, as it does for $\lambda\phi^4$ in 4D, it is *marginally irrelevant*. Finally, we see that higher-dimension operators run according to a power law, Λ^{n-d}, and these are called *irrelevant*. The term is used in a technical, not a literal sense. One might find it surprising that a diagram like Fig. 9.6 which contributes to the 6-point function is considered to be irrelevant in the IR—after all, the propagator for the internal line gets *bigger* at low energies! However, we are assuming there is a mass in the theory, so when we go really deeply into the IR, the propagator can no longer grow.

From this discussion we can get more insight into our calculation of the Wilson-Fisher fixed point (7.23). We only get the interesting term $-2\epsilon\lambda$ because we defined λ to be dimensionless. This is precisely what we have done to all the couplings when we went to the dimensionless momentum variable in Wilson's approach. We are taking the running cutoff to be the unit of energy—the renormalization scale played the same role in the previous calculation. It should now be clear that the cutoff in Wilson's language is really the same thing as what we called the renormalization scale in our first approach to renormalization.

We have touched only briefly on the profound implications of Wilson's viewpoint. One of the important concepts which is closely tied to his work is that of *universality*. As a result of the irrelevant nature of higher dimensional operators, we can modify the Lagrangian at high scales in an infinite variety of ways without affecting the low energy physics at all, provided the system approaches the IR fixed point. The microscopic Lagrangian may look very different, yet the IR limit is insensitive to this. Any two theories which flow to the same IR fixed point are said to belong to the same *universality class*.

Before leaving Wilson's contributions, let us note the relation between the fixed point and critical phenomena, namely second order phase transitions [7]. Recall the scaling behavior of renormalized amplitudes with energy,

$$G_n^{(r)}(E, \theta_i, \lambda(\mu), m(\mu), \mu) = E^D \left[\frac{Z(E)}{Z(\mu)} \right]^{n/2} f(\theta_i, \lambda(E), m^2(E)/E^2), \quad (9.47)$$

where θ_i are dimensionless kinematic variables, and D is the dimension of the operator giving rise to the n-point amplitude. Near the fixed point, we can write

$$m^2(\mu) \sim C\mu^{2\gamma_m^*}; \qquad Z^{1/2}(\mu) \sim C'\mu^{\gamma^*}. \qquad (9.48)$$

Therefore

$$G_n^{(r)} \sim E^{D+n\gamma^*} f(CE^{2(\gamma_m^*-1)}). \qquad (9.49)$$

In a second order phase transition, the correlation length ξ diverges:

$$\xi^{-1} = \lim_{r \to \infty} \frac{G_2(r)'}{G_2(r)} = \lim_{r \to \infty} \frac{(e^{-mr})'}{e^{-mr}} \doteq m, \qquad (9.50)$$

That is, the mass goes to zero. We see that as long as $\gamma_m^* > 0$, the behavior of $m(\mu)$ in (9.48) is such that $m \to 0$ as $\mu \to 0$, corroborating our claim that the fixed point corresponds to a phase transition. In statistical physics, the correlation length diverges as a power of the deviation of the temperature from its critical value in a second order phase transition:

$$\xi \sim (T - T_c)^{-\nu}. \qquad (9.51)$$

The relation between the critical exponent ν and γ_m^* is obtained from rewriting the amplitude in terms of distance scales $L \sim 1/E$,

$$G_n^{(r)} \sim L^{-D-n\gamma^*} f(m^2(L)L^2), \qquad (9.52)$$

and the fact that m^2 vanishes analytically with $(T - T_c)$:

$$C \sim T - T_c. \qquad (9.53)$$

Therefore

$$f = f((T - T_c)L^{2-2\gamma_m^*}) = f(L/\xi), \qquad (9.54)$$

which means that

$$\xi \sim (T - T_c)^{1/(2-2\gamma_m^*)}, \qquad (9.55)$$

and the critical exponent is given by

$$\nu^{-1} = 2 - 2\gamma_m^* = 2 - \epsilon/3 + O(\epsilon^2). \qquad (9.56)$$

This exponent has been computed to order ϵ^3 in an expansion in powers of ϵ, which is then evaluated at $\epsilon = 1$, corresponding to 3 dimensions (where now $d = 4 - \epsilon$). This expansion gives surprisingly good results; the Ising model (in the same universality class as ϕ^4 theory) is measured to have $\nu = 0.642 \pm 0.003$, while the ϵ expansion gives 0.626. There are other critical exponents as well; the power of L in front of the two-point function is

$$\eta = 2\gamma^* = \frac{\epsilon^2}{54} + O(\epsilon^3), \tag{9.57}$$

which gives 0.037 in the ϵ expansion, compared to 0.055 ± 0.010 for the Ising model.

Chapter 10
Fermions

Up to now we have been dealing with scalars only, but in the real world all known matter particles (as opposed to gauge bosons) are fermions. Let's recall the Lagrangians for spin 1/2 particles:

$$\text{left-handed Weyl:} \quad \psi_L^\dagger(i\bar{\sigma}\cdot\partial)\psi_L \tag{10.1}$$

$$\text{right-handed Weyl:} \quad \psi_R^\dagger(i\sigma\cdot\partial)\psi_R \tag{10.2}$$

$$\text{Dirac:} \quad \bar{\psi}(i\slashed{\partial} - m)\psi \tag{10.3}$$

$$\text{Majorana:} \quad \frac{1}{2}\bar{\psi}(i\slashed{\partial} - m)\psi \tag{10.4}$$

where $\sigma^\mu = (1, \sigma^i)$, and $\bar{\sigma}^\mu = (1, -\sigma^i)$. Weyl spinors are two component. Recall why the equation of motion $i\bar{\sigma}\cdot\partial\,\psi = 0$ implies a left-handed fermion: we can write it as

$$i(\bar{\sigma}^\mu\partial_\mu)e^{-ip_\nu x^\nu}u(p) = (\bar{\sigma}^\mu p_\mu)e^{-ip_\nu x^\nu}u = (E - \sigma^i p_i)e^{-ip_\nu x^\nu}u$$
$$= (E + \sigma^i p^i)e^{-ip_\nu x^\nu}u = (E + \vec{\sigma}\cdot\vec{p})e^{-ip_\nu x^\nu}u . \tag{10.5}$$

Therefore, since $|\vec{p}| = E$ for a massless particle,

$$\vec{\sigma}\cdot\hat{p}\,u(p) = -\frac{E}{|\vec{p}|}u(p) = -u(p), \tag{10.6}$$

hence $u(p)$ is an eigenstate of helicity with eigenvalue -1. In the massless limit, helicity and chirality coincide, and the spin being anti-aligned with the momentum is what we mean by left-handed.

Dirac spinors are four-component, made from two Weyl spinors by

$$\psi = \begin{pmatrix} \psi_R \\ \psi_L \end{pmatrix} \tag{10.7}$$

The original version of this chapter was revised: The errors in this chapter have been corrected. The correction to this chapter can be found at https://doi.org/10.1007/978-3-030-56168-0_16

J. M. Cline, *Advanced Concepts in Quantum Field Theory*, SpringerBriefs in Physics, https://doi.org/10.1007/978-3-030-56168-0_10

using the chiral representation of the gamma matrices, where

$$\gamma^0 = \begin{pmatrix} 0 & 1 \\ 1 & 0 \end{pmatrix}; \qquad \gamma^i = \begin{pmatrix} 0 & -\sigma^i \\ \sigma^i & 0 \end{pmatrix}; \qquad \gamma_5 = \begin{pmatrix} 1 & 0 \\ 0 & -1 \end{pmatrix}; \qquad (10.8)$$

they obey the anticommutation relations

$$\{\gamma^\mu, \gamma^\nu\} = 2g^{\mu\nu}. \qquad (10.9)$$

A Majorana spinor is made from a single Weyl spinor using the fact that the charge conjugate of ψ_L, $\sigma_2 \psi_L^*$, behaves in the same way as ψ_R under Lorentz transformations:

$$\psi = \begin{pmatrix} \sigma_2 \psi_L^* \\ \psi_L \end{pmatrix}. \qquad (10.10)$$

If one writes out the Majorana Lagrangian in terms of two-component spinors, it becomes clear why the factor of $\frac{1}{2}$ is needed:

$$\mathcal{L}_{\text{Dirac}} = \psi_L^\dagger (i\bar{\sigma} \cdot \partial)\psi_L + \psi_R^\dagger (i\sigma \cdot \partial)\psi_R - m\psi_L^\dagger \psi_R - m\psi_R^\dagger \psi_L, \qquad (10.11)$$

whereas for the Majorana case, the term coming from $\psi_R^\dagger (i\sigma \cdot \partial)\psi_R$ would be a double-counting of $\psi_L^\dagger (i\bar{\sigma} \cdot \partial)\psi_L$ without the extra factor of $\frac{1}{2}$.

You have already learned that the Feynman propagator for a Dirac fermion is

$$S_F(x', x)_{\beta\alpha} = -i\langle 0|T(\psi_\beta(x')\bar{\psi}_\alpha(x))|0\rangle \qquad (10.12)$$

$$= \int \frac{d^4 p}{(2\pi)^4} \frac{e^{-ip\cdot(x'-x)}}{\not{p} - m + i\varepsilon} = \int \frac{d^4 p}{(2\pi)^4} e^{-ip\cdot(x'-x)} \frac{\not{p} + m}{p^2 - m^2 + i\varepsilon}. \qquad (10.13)$$

For Weyl fermions, we can write similar expresssions, for example

$$S_F(x', x)_{\beta\alpha} = -i\langle 0|T(\psi_\beta(x')\bar{\psi}_\alpha(x))|0\rangle \qquad (10.14)$$

$$= \int \frac{d^4 p}{(2\pi)^4} \frac{e^{-ip\cdot(x'-x)}}{\sigma \cdot p + i\varepsilon} = \int \frac{d^4 p}{(2\pi)^4} e^{-ip\cdot(x'-x)} \frac{\bar{\sigma} \cdot p}{p^2 + i\varepsilon} \qquad (10.15)$$

Here I used the identity $\sigma \cdot p \, \bar{\sigma} \cdot p = p^2$. However, it is usually more convenient to work in terms of four-component fields using the Majorana or Dirac form of the Lagrangian. In this way we can always use the familiar Dirac algebra instead of having to deal with the 2 × 2 sigma matrices. For massless neutrinos, there is no harm in pretending there is an additional right-handed neutrino in the theory, as long as it has no interactions with real particles. This is why we can use the Dirac propagator in calculations where the neutrino mass can be ignored. If we are considering a Weyl field which has a mass, one must use the Majorana form of the Lagrangian. In this case one has to be more careful when computing Feynman diagrams because there are more ways of Wick-contracting the fermion fields, as I shall explain below.

Note: there is another kind of mass term one can write down for Dirac or Majorana fermions:

$$m\bar{\psi}\psi + im_5\bar{\psi}\gamma_5\psi. \tag{10.16}$$

To solve $(\not{p} - m - im_5\gamma_5)u(p) = 0$, consider the form $u(p) = (\not{p} + m - im_5\gamma_5)v(p)$ where $v(p)$ is an arbitrary spinor. The equation of motion gives

$$p^2 - m^2 - m_5^2 = 0, \tag{10.17}$$

so $m^2 + m_5^2$ is the complete mass squared. In fact it is possible to tranform the usual Dirac mass term into this form by doing a field redefinition

$$\psi \to e^{i\theta\gamma_5}\psi = (\cos\theta + i\sin\theta\gamma_5)\psi. \tag{10.18}$$

Then

$$m \to e^{i\theta\gamma_5}me^{i\theta\gamma_5} = (\cos 2\theta + i\sin 2\theta\gamma_5)m, \tag{10.19}$$

showing that one can freely transform between the usual form of the mass and the parity-violating form. Of course, this will also change the form of the interactions of ψ, so it is usually most convenient to keep the mass in the usual form.

In analogy with bosons, the Feynman rule for a fermion line is given by

$$\xrightarrow{\;\;p\;\;} = \frac{i}{\not{p} - m + i\varepsilon}. \tag{10.20}$$

Unlike bosons, fermions have a directionality attached to their lines because of the fact that the propagator is not symmetric under $p \to -p$. So we need to pay attention to the direction of flow of momentum.

In addition to the rule for internal lines, we need to include spinors for the external lines. From the canonical formalism, it is straightforward to derive:

$$
\begin{aligned}
&u(p, s) &&\text{for each particle entering graph}\\
&v(p, s) &&\text{for each antiparticle entering graph}\\
&\bar{u}(p, s) &&\text{for each particle leaving graph}\\
&\bar{v}(p, s) &&\text{for each antiparticle leaving graph}.
\end{aligned} \tag{10.21}
$$

As an example of using these rules, let's compute the cross section for charged-current scattering of a neutrino on a neutron into electron plus proton. At low energies, the interaction Lagrangian has the form

$$\sqrt{2}G_F[\bar{p}(c_v - c_a\gamma_5)\gamma_\mu n][\bar{e}\gamma^\mu P_L\nu] + \text{h.c.}, \tag{10.22}$$

where the Fermi constant G_F arises from the internal propagator of the W boson, which we have integrated out to obtain the nonrenormalizable effective interaction

(10.22). The nucleon vector and axial vector couplings are given by $c_v = 1$ (which can be derived on theoretical grounds) and $c_a = -1.25$, which is experimentally measured.

The amplitude for $\nu n \to ep$ is

$$\mathcal{M} = \sqrt{2} G_F \left[\bar{u}_p(c_v - c_a\gamma_5)\gamma_\mu u_n\right]\left[\bar{u}_e\gamma^\mu P_L u_\nu\right], \qquad (10.23)$$

and the amplitude squared is

$$|\mathcal{M}|^2 = 2G_F^2 \left[\bar{u}_p(c_v - c_a\gamma_5)\gamma_\mu u_n\right]\left[\bar{u}_e\gamma^\alpha P_L u_\nu\right]\left[\bar{u}_n(c_v - c_a\gamma_5)\gamma_\alpha u_p\right]\left[\bar{u}_\nu\gamma^\mu P_L u_e\right]. \qquad (10.24)$$

To get this, one uses the result that

$$(\bar{u}_1 X u_2)^* = u_2^\dagger X^\dagger \gamma_0^\dagger u_1 \qquad (10.25)$$

and

$$\gamma^{\mu\dagger}\gamma^{0\dagger} = \gamma^0\gamma^\mu; \qquad \gamma_5^\dagger\gamma^{0\dagger} = -\gamma^0\gamma_5. \qquad (10.26)$$

Next we sum over final polarizations and average over initial ones using the formulas

$$\sum_s u(p,s)\bar{u}(p,s) = \not{p} + m; \qquad \sum_s v(p,s)\bar{v}(p,s) = \not{p} - m, \qquad (10.27)$$

to obtain

$$\langle|\mathcal{M}|^2\rangle = \frac{1}{4}\sum|\mathcal{M}|^2 = \frac{1}{2}G_F^2 \,\text{tr}\left[P_L(\not{p}_e + m_e)\gamma^\alpha P_L\not{p}_\nu\gamma^\mu\right]\cdot\text{tr}\left[(\not{p}_n + m_n)X\gamma_\alpha(\not{p}_p + m_p)X\gamma_\mu\right], \qquad (10.28)$$

where now $X = c_v - c_a\gamma_5$. Thus, ignoring the electron mass,

$$\begin{aligned}
\langle|\mathcal{M}|^2\rangle &= \frac{1}{2}G_F^2 \,\text{tr}\left[\not{p}_e\gamma^\alpha\not{p}_\nu\gamma^\mu\frac{1}{2}(1+\gamma_5)\right]\cdot\left(\text{tr}\left[\not{p}_n X\gamma_\alpha\not{p}_p X\gamma_\mu\right] + m_n m_p\text{tr}\left[X\gamma_\alpha X\gamma_\mu\right]\right) \\
&= \frac{4}{4}G_F^2\left(p_e^\mu p_\nu^\alpha + p_e^\alpha p_\nu^\mu - p_e\cdot p_\nu\eta^{\mu\alpha} + i\epsilon^{\rho\mu\sigma\alpha}p_{e,\rho}p_{\nu,\sigma}\right) \\
&\quad \cdot\left(\text{tr}\left[\gamma_\alpha\not{p}_n\gamma_\mu\not{p}_p(c_v - c_a\gamma_5)^2\right] + m_n m_p tr\left[\gamma_\mu(c_v + c_a\gamma_5)(c_v - c_a\gamma_5)\gamma_\alpha\right]\right) \\
&= \frac{16}{4}G_F^2\left(p_e^\mu p_\nu^\alpha + p_e^\alpha p_\nu^\mu - p_e\cdot p_\nu\eta^{\mu\alpha} + i\epsilon^{\rho\mu\sigma\alpha}p_{e\rho}p_{\nu\sigma}\right) \\
&\quad \left([p_{n,\alpha}p_{p,\mu} + p_{p,\alpha}p_{n,\mu}] - p_n\cdot p_p\eta_{\alpha\mu}](c_v^2 + c_a^2)\right. \\
&\quad \left. - 2ic_vc_a\epsilon_{\alpha\rho'\mu\sigma'}p_n^{\rho'}p_p^{\sigma'} + m_n m_p(c_v^2 - c_a^2)\eta_{\alpha\mu}\right) \\
&= 4G_F^2\left(2(p_e\cdot p_p\,p_\nu\cdot p_n + p_e\cdot p_n\,p_\nu\cdot p_p)(c_v^2 + c_a^2) + m_n m_p(c_v^2 - c_a^2)(2p_e\cdot p_\nu - 4p_e\cdot p_\nu)\right. \\
&\quad \left. + 2c_vc_a(2\delta_{\rho'\sigma'}^{\rho\sigma})p_{e,\rho}p_{\nu,\sigma}p_n^{\rho'}p_p^{\sigma'}\right) \\
&= 8G_F^2\left(p_e\cdot p_p\,p_\nu\cdot p_n(c_v - c_a)^2 + p_e\cdot p_n\,p_\nu\cdot p_p(c_v + c_a)^2 - m_n m_p\,p_e\cdot p_\nu(c_v^2 - c_a^2)\right).
\end{aligned}$$

$$(10.29)$$

Note that I used the identity $\epsilon_{\alpha\rho'\mu\sigma'}\epsilon^{\alpha\rho\mu\sigma} = 2\delta^{\rho\sigma}_{\rho'\sigma'}$. To simplify this, let's go to the low energy limit where the proton and neutron are at rest, and the neutrino energy is much smaller than the nucleon mass. Then

$$\langle|\mathcal{M}|^2\rangle \cong 8G_F^2 m_n m_p \left(2E_e E_\nu(c_v^2 + c_a^2) - p_e \cdot p_\nu(c_v^2 - c_a^2)\right)$$

$$\cong \frac{8}{16}G_F^2 m_n m_p(91 E_e E_\nu - 9\vec{p}_e \cdot \vec{p}_\nu). \tag{10.30}$$

Recall that the differential cross section is given by

$$\frac{d\sigma}{dt} = \frac{1}{64\pi s}\frac{1}{|p_{1,\text{cm}}|^2}\langle|\mathcal{M}|^2\rangle, \tag{10.31}$$

where $p_{1,\text{cm}} = p_{1,\text{lab}}m_n/\sqrt{s} = E_\nu m_n/\sqrt{s}$. Ignoring the small $\vec{p}_e \cdot \vec{p}_\nu$ term, we can then estimate

$$\frac{d\sigma}{dt} \cong \frac{1}{64\pi}\frac{1}{E_\nu^2 m_n^2}\langle|\mathcal{M}|^2\rangle$$

$$\cong \frac{91}{2}G_F^2 m_n m_p E_\nu E_e$$

$$\cong \frac{46}{64\pi}G_F^2. \tag{10.32}$$

We used that fact that, by energy conservation, since $E_\nu \ll m_n$, $E_e \cong E_\nu$. The maximum momentum transfer is $\sqrt{-t} = 2E_\nu$, when the electron comes out in the opposite direction to the neutrino, thus the total scattering cross section is

$$\sigma = \int_{-(2E_n u)^2}^{0} \frac{d\sigma}{dt} \cong \frac{46}{16\pi}G_F^2 E_\nu^2. \tag{10.33}$$

Now that we have reviewed fermions at tree level, we can look at how they behave in loop diagrams. For this purpose we would like to introduce interactions. The 4-fermion Lagrangian we just studied is not renormalizable, since by power counting diagrams like those in Fig. 10.1 are divergent. On the other hand, interactions with scalars can be renormalizable:

$$\mu^\epsilon \phi \bar{\psi}(g_1 + ig_2\gamma_5)\psi. \tag{10.34}$$

These are called Yukawa couplings; g_1 is called a scalar coupling and g_2 is a pseudoscalar coupling, meaning that under parity, if ϕ was even and $g_2 = 0$, the interaction would be invariant, whereas if ϕ was odd and $g_1 = 0$, it would also be invariant. In the

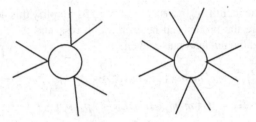

Fig. 10.1 Divergent loop diagrams in 4-fermion theory

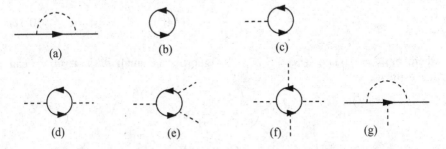

Fig. 10.2 Divergent loop diagrams with Yukawa couplings to a scalar field (dashed lines)

first case, ϕ is a scalar under parity, and in the second it is a pseudoscalar, hence the terminology. In either case, the dimension of the operator in $d = 4 - 2\epsilon$ dimensions requires the factor of μ^ϵ. For simplicity, let's consider the case of the scalar coupling, and denote $g_1 = g$. Then there are a few divergent diagrams we can consider, shown in Fig. 10.2.

At this point I will leave detailed computations for you to derive in the exercises, and just give an outline of the calculation. We start with Fig. 10.2d, the correction to the scalar inverse propagator. The important points are the following. (1) the fermion loop gets an extra minus sign relative to boson loops; this occurs because one has to anticommute one pair of fermion fields in order for them to be in the right order to give the propagator when the Wick contractions are done. (2) There is a trace over the Dirac indices associated with each fermion loop. When we use dimensional regularization, we have to know how Dirac matrices generalize to higher dimensions. Here are some relevant formulas.

$$\gamma_\mu\gamma^\mu = d \tag{10.35}$$

$$\operatorname{tr} 1 = d \tag{10.36}$$

$$\operatorname{tr}\gamma_\mu\gamma_\alpha = d\eta_{\mu\alpha} \tag{10.37}$$

$$\operatorname{tr}\gamma_\mu\gamma_\alpha\gamma_\nu\gamma_\beta = d(\eta_{\mu\alpha}\eta_{\nu\beta} + \eta_{\mu\beta}\eta_{\nu\alpha} - \eta_{\mu\nu}\eta_{\alpha\beta} \tag{10.38}$$

$$\gamma_\mu\gamma_\alpha\gamma^\mu = -\gamma_\mu\gamma^\mu\gamma_\alpha + 2\gamma_\mu\delta^\mu_\alpha = (2 - d)\gamma_\alpha \tag{10.39}$$

$$\gamma_\mu\slashed{a}\slashed{b}\gamma^\mu = 4a\cdot b + (d - 4)\slashed{a}\slashed{b} \tag{10.40}$$

$$\gamma_\mu \not{a} \not{b} \not{c} \gamma^\mu = -2\not{c} \not{b} \not{a} + (4-d)\not{a} \not{b} \not{c} \tag{10.41}$$

As for formulas involving γ_5, there is no natural way to continue them to other dimensions, so we are forced to use the 4D formulas. This turns to out be consistent for many purposes, but we will see later that it leads to some problems in rigorously defining a theory with chiral fermions with gauge interactions.

From the result of evaluating the diagram of Fig. 10.2d, we can verify that the divergent part of the correction to the mass has the opposite sign to what we found in ϕ^4 theory:

$$\delta m^2_{\text{scalar}} = \frac{g^2}{16\pi^2} \frac{12}{\epsilon} m^2_{\text{fermion}} . \tag{10.42}$$

In fact the divergences of the fermion and scalar loops cancel each other if the Yukawa and quartic couplings are related by

$$g^2 = \frac{\lambda}{4!} , \tag{10.43}$$

and if the masses of the fermion and scalar are equal to each other. If there was no symmetry in the theory to enforce this kind of relationship, these would be a kind of fine-tuning. However, supersymmetry (SUSY) enforces precisely this kind of relationship. In SUSY theories, this kind of cancellation also leaves the corrections to the scalar quartic coupling finite; notice that Fig. 10.2f goes like g^4/ϵ, whereas Fig. 2.4 (the scalar loop contribution to the quartic coupling) goes like λ^2/ϵ, so these do have the correct form to cancel each other.

On the other hand if we look at wave function renormalization, Fig. 10.2f gives

$$-\!\!-\!\!\bigcirc\!\!-\!\!- \sim i \frac{g^2}{16\pi^2} \frac{2}{\epsilon} p^2 \tag{10.44}$$

$$[\text{compare to} \quad -\!\!-\!\!-\!\!- = i(p^2 - m^2) \,] , \tag{10.45}$$

whereas we know that the setting sun diagram of scalar field theory goes like λ^2/ϵ. Thus these two diagrams do not have the right form to cancel each other. This shows that while SUSY can prevent divergences in masses and couplings, it does not do so for wave function renormalization.

Another consequence of SUSY is that the vacuum diagrams of the form $\bigcirc$ must cancel. We would like to see how Fig. 10.2b gets the opposite sign to that of the scalar loop. This comes about because of the properties of the fermionic form of the path integral:

$$\bigcirc = \Gamma = \frac{1}{i} \ln \left(\int \mathcal{D}\psi \mathcal{D}\bar{\psi} e^{iS} \right) . \tag{10.46}$$

Consider the Grassmann integral

$$\int \prod_i d\eta_i^* \, d\eta_i \, e^{i\eta_i^* M_{ij}\eta_j} = \det M \,. \tag{10.47}$$

This is in contrast to the bosonic path integral which gives $(\det M)^{-1/2}$. The important thing here is the sign of the exponent. When we take the log, this explains why the fermionic loop has the opposite sign to the bosonic one. As for the factor of $1/2$ which we had for bosons, this was due to the fact that a real scalar field is just one degree of freedom, whereas a Weyl fermion has two spin components, and therefore is two degrees of freedom. If we consider the path integral over a complex scalar field $\phi = (\phi_1 + i\phi_2)/\sqrt{2}$, we would obtain $(\det M)^{-1}$. We now do a detailed computation of this determinant for fermions to show that indeed the cancellation is exact between a Weyl fermion and a complex scalar if they have the same mass, or between a Dirac fermion and two such complex scalars. In SUSY theories, scalars are always complex, and their corresponding fermionic partners are always Weyl.

Just like for bosons, we can construct the Coleman-Weinberg potential for a scalar field due to the fermion loop, and the result is the fermion determinant with the mass of the fermion evaluated using the background scalar field:

$$V_{CW} = -\frac{N}{64\pi^2}(m + g\phi)^4 \ln\left(\frac{(m + g\phi)^2}{\mu^2}\right), \tag{10.48}$$

where $N = 4$ for a Dirac and 2 for a Majorana or Weyl fermion. This follows from the result that

$$\Gamma = -i \ln \det(i\partial\!\!\!/ - m) = -i\frac{N}{2}\int d^4x \int \frac{d^4p}{(2\pi)^4} \ln(p^2 - m^2) \,. \tag{10.49}$$

Notice that when we Wick-rotate, Γ gets a positive contribution from fermions. This means the vacuum energy density gets a negative contribution, since action $\sim (-\text{ potential})$.

We return to the diagrams of Fig. 10.2. The tadpole (c) is similar to that which arises in ϕ^3 theory and must be renormalized similarly, by introducing a counterterm which is linear in ϕ. Let us focus on diagram (a) instead; we call this a contribution to the self-energy of the fermion. To help get the sign of this contribution right, I will use the background field method. This gives

$$i\Gamma_{1\text{loop}} = \frac{1}{2!}(-ig\mu^\epsilon)^2 \left\langle \left(\int d^4x \, (\bar{\psi}_c\phi_q\psi_q + \bar{\psi}_q\phi_q\psi_c) \int d^4y \, (\bar{\psi}_c\phi_q\psi_q + \bar{\psi}_q\phi_q\psi_c) \right) \right\rangle$$

$$= (-ig\mu^\epsilon)^2 \int d^4x \int d^4y \, \bar{\psi}_c(x) \langle \phi_q(x)\phi_q(y) \rangle \langle \psi_q(x)\bar{\psi}_q(y) \rangle \psi_c(y)$$

$$= -g^2\mu^{2\epsilon} \int \frac{d^4p}{(2\pi)^4} \bar{\psi}_c(p) \int \frac{d^4q}{(2\pi)^4} \frac{i}{p\!\!\!/ + q\!\!\!/ - m_f} \frac{i}{q^2 - m_s^2} \psi_c(p) \,. \tag{10.50}$$

since there are two ways of contracting the quantum fermion fields. This expression is to be compared to the tree-level one,

$$\Gamma_{\text{tree}} = \int \frac{d^4 p}{(2\pi)^4} \bar{\psi}_c(p)(\not{p} - m_f)\psi_c(p).$$ (10.51)

We can in this way identify the corrections to the mass and wave function renormalization. The correction to the fermion inverse propagator is often called the *self-energy*, $\Sigma(p)$. The full inverse propagator is then $\not{p} - m + \Sigma$. Comparing the above equations, we see that

$$\Sigma = ig^2\mu^{2\epsilon} \int \frac{d^4 q}{(2\pi)^4} \frac{\not{p} + \not{q} + m_f}{(p+q)^2 - m_f^2} \frac{1}{q^2 - m_s^2}.$$ (10.52)

Evaluating it in the usual way, we find the result

$$\Sigma = g^2\mu^{2\epsilon} \int\limits_0^1 dx \frac{\Gamma(\epsilon)}{(4\pi)^{2-\epsilon}\Gamma(2)} \frac{\not{p}(1-x) + m_f}{(M^2)^\epsilon},$$ (10.53)

where $M^2 = xm_f^2 + (1-x)m_s^2 - p^2 x(1-x)$. The most divergent part is

$$\underset{\text{───────}}{\overset{\overparen{\quad\quad}}{}} = \Sigma \sim \frac{g^2}{16\pi^2\epsilon}\left(\frac{1}{2}\not{p} + m_f\right).$$ (10.54)

Finally, let's evaluate the remaining divergent diagram, Fig. 10.2g, which is called the *vertex correction*. To simplify things, let's evaluate it at vanishing external momenta, since this is all we need for getting the divergent contribution to the Yukawa coupling. Evaluating the diagram directly, we find the combinatoric factor is 1, hence

$$
\begin{aligned}
i\Gamma_{\text{vertex}} &= (-ig\mu^\epsilon)^3 \int \frac{d^4 q}{(2\pi)^4} \frac{i}{\not{q} - m_f} \frac{i}{\not{q} - m_f} \frac{i}{q^2 - m_s^2} \\
&= (g\mu^\epsilon)^3 \int \frac{d^4 q}{(2\pi)^4} \frac{(\not{q} + m_f)^2}{(q^2 - m_f^2)^2(q^2 - m_s^2)} \\
&= (g\mu^\epsilon)^3 \int \frac{d^4 q}{(2\pi)^4} \left(\frac{1}{(q^2 - m_f^2)(q^2 - m_s^2)} + \text{convergent piece}\right)
\end{aligned}
$$

$$= i(g\mu^\epsilon)^3 \int\limits_0^1 dx \, \frac{\Gamma(\epsilon)}{(4\pi)^{2-\epsilon}(M^2)^\epsilon} \qquad (M^2 \equiv xm_f^2 + (1-x)m_s^2)$$

$$\sim i \frac{(g\mu^\epsilon)^3}{16\pi^2\epsilon}, \tag{10.55}$$

which should be compared to the tree level result $i\Gamma = -ig\mu^\epsilon$.

Now we can put together the above results to do the renormalization of this theory and find the beta function for the Yukawa coupling. Let us write the tree plus 1-loop, unrenormalized effective action in the form

$$\Gamma = \int \frac{d^4p}{(2\pi)^4} \left[\frac{1}{2}\phi_{-p} \left(p^2 - m_s^2 + \frac{a}{\epsilon}p^2 - \frac{b}{\epsilon}m_s^2 \right) \phi_p + \bar{\psi}_p \left(\slashed{p} - m_f + \frac{c}{\epsilon}\slashed{p} - \frac{d}{\epsilon}m_f \right) \psi_p \right]$$
$$- g\left(1 + \frac{e}{\epsilon}\right) \int d^4x \, \bar{\psi}\phi\psi. \tag{10.56}$$

Comparing with our previous results, and defining $\hat{g} = g\mu^\epsilon/4\pi$, we can identify

$$a = 2\hat{g}^2; \quad b = 12\frac{m_f^2}{m_s^2}\hat{g}^2; \quad c = \frac{1}{2}\hat{g}^2; \quad d = -\hat{g}^2; \quad e = -\hat{g}^2. \tag{10.57}$$

The first step is to renormalize the fields to absorb the a and c terms. We define

$$\psi = \sqrt{Z_\psi}\psi_r \cong \left(1 - \frac{c}{2\epsilon}\right)\psi_r; \quad \phi = \sqrt{Z_\phi}\phi_r \cong \left(1 - \frac{a}{2\epsilon}\right)\phi_r. \tag{10.58}$$

The effective action becomes

$$\Gamma = \int \frac{d^4p}{(2\pi)^4} \left[\frac{1}{2}\phi_{r,-p}\left(p^2 - m_s^2\left(1 + \frac{b-a}{\epsilon}\right)\right)\phi_{r,p} + \bar{\psi}_{r,p}\left(\slashed{p} - m_f\left(1 + \frac{d-c}{\epsilon}\right)\right)\psi_{r,p} \right]$$
$$- g\mu^\epsilon\left(1 + \frac{e-c-a/2}{\epsilon}\right)\int d^4x \, \bar{\psi}_r\phi_r\psi_r. \tag{10.59}$$

It is now clear how to define the bare parameters in order to absorb the divergences:

$$m_{s,0}^2 = \left(1 - \frac{b-a}{\epsilon}\right)m_{s,r}^2; \quad m_{f,0} = \left(1 - \frac{d-c}{\epsilon}\right)m_{f,r}; \quad g_0 = \left(1 - \frac{e-c-a/2}{\epsilon}\right)g_r. \tag{10.60}$$

Using the determined values for e, c and a, and remembering that g_0 has dimensions given by μ^ϵ, the full bare coupling is (see the parallel discussion starting at (7.19) for the qruartic scalar coupling)

$$g_{\text{bare}} = \mu^\epsilon \left(1 + \frac{5\hat{g}_r^2}{2\epsilon}\right)g_r \equiv \mu^\epsilon \left(g + \frac{a_1(g)}{\epsilon}\right). \tag{10.61}$$

Now we can compute the β function for the Yukawa coupling,

$$0 = \mu\frac{\partial}{\partial\mu}g_{\text{bare}} = \epsilon\left(g + \frac{a_1(g)}{\epsilon}\right) + \beta(g)\left(1 + \frac{a_1'}{\epsilon}\right)$$

$$\longrightarrow \quad \beta(g) \cong \epsilon\left(g + \frac{a_1(g)}{\epsilon}\right)\left(1 - \frac{a_1'}{\epsilon}\right)$$

$$= \epsilon g + \frac{5g^3}{16\pi^2}. \tag{10.62}$$

Thus, like the quartic coupling, the Yukawa coupling is not asymptotically free; it flows to larger values in the ultraviolet.

I must confess that this way of computing the beta function from dimensional regularization seems rather abstract and unintuitive to me. Here is another way which is more down-to-earth. Go back to the 1-loop effective action (10.59) which has been wave-function-renormalized, and imagine what the 3-point function would look like if we had computed it using a momentum-space cutoff, and at nonvanishing external momentum. Remembering that $1/\epsilon$ corresponds to $\ln\Lambda^2$, and taking the external momentum to be p, the effective coupling would have the form

$$g_{\text{eff}} = g\left(1 - \frac{5g^2}{32\pi^2}\ln(\Lambda^2/p^2)\right). \tag{10.63}$$

When we renormalize, this will become

$$g_{\text{eff}} = g(\mu)\left(1 - \frac{5g^2(\mu)}{32\pi^2}\ln(\mu^2/p^2)\right). \tag{10.64}$$

Notice that the energy-dependent coupling increases with p^2 as expected from the sign of the beta function. We can now compute the beta function by demanding that the physical coupling g_{eff} be independent of μ:

$$\mu\frac{\partial}{\partial\mu}g_{\text{eff}} = 0 = \beta\left(1 + O(g^2)\right) - \frac{5g^3}{16\pi^2}, \tag{10.65}$$

which gives the same result as DR to the order we are computing. You can see the parallels between this way of doing it and the DR method. I tend to think that the DR procedure is a set of manipulations that are mathematically equivalent to using a momentum space cutoff (apart from issues of gauge invariance), but the latter gives a more concrete way of understanding what we are actually doing.

To conclude this chapter about fermions, I want to discuss the differences between Majorana and Dirac particles. The previous computation for the beta function was for a Dirac fermion. Implicit in this discussion was the fact that only Green's functions like $\langle\psi\bar\psi\rangle$ were nonzero, whereas $\langle\psi\psi\rangle$ was assumed to be zero. Here I would like to show that this is true for Dirac fermions, but not Majorana ones. To put the discussion in context, let's suppose that we are doing the Wick contractions for a combination of fields like

$$\langle (\bar{\psi}X\psi)_x (\bar{\psi}Y\psi)_y \rangle_{\text{con}} , \qquad (10.66)$$

where X and Y are matrices such as 1, γ_μ, γ_5, $\gamma_\mu\gamma_5$, *etc.*, and we are only interested in the connected contributions. As we have noted before, one possible contraction gives

$$-\operatorname{tr}(X\langle\psi_x\bar{\psi}_y\rangle Y\langle\psi_y\bar{\psi}_x\rangle) . \qquad (10.67)$$

To write the other possible contraction involving $\langle\psi_x\psi_y\rangle\langle\bar{\psi}_x\bar{\psi}_y\rangle$ in a convenient way, let's first rewrite (10.66) by making use of the fact that

$$\begin{aligned}
(\bar{\psi}X\psi) &= -\psi^T X^T \gamma_0 \psi^* \\
&= -\psi^T(-\gamma^{2\dagger}\gamma_0)(-\gamma^{2\dagger}\gamma_0)^{-1}X^T\gamma_0(-\gamma^2)^{-1}(-\gamma^2)\psi^* \\
&\equiv \bar{\psi}^c X^c \psi^c ,
\end{aligned} \qquad (10.68)$$

where we have defined the charge-conjugated spinor

$$\psi^c = -\gamma^2\psi^* \qquad (10.69)$$

and

$$X^c = -\gamma_0(-\gamma^{2\dagger})^{-1}X^T\gamma_0\gamma^2 = \gamma_0\gamma^2 X^T\gamma_0\gamma^2 . \qquad (10.70)$$

Notice that γ^2 is antihermitian in the representation we are using. The leading minus signs in (10.68) and (10.70) come from anticommuting the two fields. When we rewrite $(\bar{\psi}X\psi)_x$ in this way, we see that the other possible contraction of the fields takes the form

$$-\operatorname{tr}(X^c\langle\psi_x^c\bar{\psi}_y\rangle Y\langle\psi_y\bar{\psi}_x^c\rangle) . \qquad (10.71)$$

The fact that $\langle\psi_x^c\bar{\psi}_y\rangle$ vanishes for a Dirac fermion follows simply from the fact that fermion number is a conserved quantity in the Dirac Lagrangian. The latter is symmetric under the transformations

$$\psi \to e^{i\theta}\psi, \qquad \bar{\psi} \to e^{-i\theta}\bar{\psi} . \qquad (10.72)$$

However, $\psi^c \to e^{-i\theta}\psi^c$ under this symmetry, so that $\langle\psi_x^c\bar{\psi}_y\rangle \to e^{-2i\theta}\langle\psi_x^c\bar{\psi}_y\rangle$. A nonvanishing VEV for $\psi_x^c\bar{\psi}_y$ would therefore break the symmetry, in contradiction to the fact that the symmetry does exist in the underlying Lagrangian. You can see this explicitly by considering the path integral, and changing variables from ψ to $\psi' = e^{i\theta}\psi$:

$$\langle \psi_x^c \bar\psi_y \rangle = \int \mathcal{D}\psi \mathcal{D}\bar\psi e^{iS(\bar\psi,\psi)} \psi_x^c \bar\psi_y$$

$$= \int \mathcal{D}\psi' \mathcal{D}\bar\psi' e^{iS(\bar\psi',\psi')} \psi_x'^c \bar\psi_y' \qquad (\psi' = e^{i\theta}\psi)$$

$$= \int \mathcal{D}\psi \mathcal{D}\bar\psi e^{iS(\bar\psi,\psi)} \psi_x'^c \bar\psi_y'$$

$$= e^{2i\theta}\langle \psi_x^c \bar\psi_y \rangle . \tag{10.73}$$

Both the action and the path integral measure are invariant under this phase transformation. For the action this was obvious; for the measure, we note that the determinant of the Jacobian matrix is unity:

$$\left| \frac{\partial(\psi',\bar\psi')}{\partial(\psi,\bar\psi)} \right| = \left| \begin{matrix} e^{i\theta} & 0 \\ 0 & e^{-i\theta} \end{matrix} \right| = 1 . \tag{10.74}$$

Since $\langle \psi_x^c \bar\psi_y \rangle = e^{2i\theta}\langle \psi_x^c \bar\psi_y \rangle$, it must vanish.

This argument does not work for massive Majorana fermions however, since their action is not invariant under the phase transformation. More precisely, it does not make sense to perform such a phase transformation on a Majorana spinor since it is defined as

$$\psi = \begin{pmatrix} \sigma_2 \psi_L^* \\ \psi_L \end{pmatrix} , \tag{10.75}$$

which obeys the identity

$$\psi^c = \psi . \tag{10.76}$$

Clearly, a spinor which satisfied this property (being self-conjugate) would no longer do so if we multiplied it by a phase. Therefore we cannot define a symmetry operation which forbids Green's functions of the form $\langle \psi_x^c \bar\psi_y \rangle$ for Majorana fermions. In fact, since $\psi^c = \psi$, we see that $\langle \psi_x^c \bar\psi_y \rangle = \langle \psi_x \bar\psi_y \rangle$.

Now we can see how to compute Feynman diagrams with Majorana fermions. We have to compute all possible contractions of the fermionic fields, by charge-conjugating one of the bilinears in order to put terms like $\psi\psi^T$ into the more familiar form $\psi\bar\psi^c = \psi\bar\psi$. One has to know how the matrix X transforms under charge conjugation. For example, it is easy to show that

$$1^c = 1; \quad \gamma_5^c = +\gamma_5; \quad (\gamma^\mu)^c = -\gamma^\mu; \quad (\gamma_5\gamma^\mu)^c = -\gamma_5\gamma^\mu . \tag{10.77}$$

The fact that $\bar\psi\gamma_\mu\psi$ changes sign under charge conjugation is intuitively clear: for electrons this is the electric current, which should change sign when transforming from electrons to positrons.

We can now redo the previous calculations (Fig. 10.2) in a theory where the fermion is Majorana. In this case, it is convenient to normalize the interaction term with an additional factor of $\frac{1}{2}$, as we do for the kinetic term, since this preserves the relationship between the Yukawa coupling and the mass if we give a classical expectation value to the scalar field ϕ:

$$\mathcal{L} = \frac{1}{2}\bar{\psi}\left(i\partial\!\!\!/ - m - g\phi\right)\psi. \tag{10.78}$$

When we redo the diagrams of Fig. 10.2, we find that the results are identical to those of the Dirac fermion for all diagrams where the fermions are in the external states, because the new factor of $\frac{1}{2}$ in the coupling in (10.78) is exactly compensated by the number of new contractions that can be done with the fermion fields. However in the vacuum polarization Fig. 10.2d, there are only two ways of contracting the fermions, while there are two factors of $\frac{1}{2}$. This makes sense: there are only half as many fermions circulating in the loop for the Majorana case, relative to the Dirac one. Therefore the only change in computing the beta function is that the factor a in (10.59) is divided by 2:

$$\beta = \frac{4g^3}{16\pi^2} \quad \text{(Majorana case)}. \tag{10.79}$$

Chapter 11
The Axial Anomaly

Sometimes a symmetry which seems to be present in a theory at the classical level can be spoiled by quantum (loop) effects. We have already seen such an example without knowing it. A theory with no massive parameters such as

$$S = \int d^4x \left(\frac{1}{2}(\partial\phi)^2 + \bar{\psi}\partial\!\!\!/\psi - \frac{\lambda}{4!} - g\bar{\psi}\phi\psi \right) \tag{11.1}$$

has a dilatation or scale transformation symmetry

$$\phi \to e^a \phi, \quad \psi \to e^{3a/2}\psi, \quad x^\mu \to e^{-a}x^\mu, \tag{11.2}$$

since there is no mass scale in the Lagrangian. However, the Coleman-Weinberg potential breaks this symmetry due to factors like $\ln(\lambda\phi^2/\mu^2)$ and $\ln(g^2\phi^2/\mu^2)$ which are introduced through renormalization. It is impossible to avoid introducing the mass scale μ. A symmetry broken by quantum effects is said to be anomalous.

When we introduced dimensional regularization, one of our motivations was that it was important to try to preserve the symmetries which are present in the theory. In the previous section, we noted an example of one of these symmetries, the transformation $\psi \to e^{i\theta}\psi$ for Dirac fermions. Recall that for any symmetry of a Lagrangian, there is a conserved (Nöther) current which is constructed by promoting θ from a global constant to a locally varying function $\theta(x)$. The Nöther procedure is to find the variation of the action under an infinitesimal transformation of this kind. If it is a symmetry, then the result can always be written in the form

$$\delta\mathcal{L} = \int d^4x \, \partial_\mu\theta J^\mu = -\int d^4x \, \theta\partial_\mu J^\mu = 0, \tag{11.3}$$

where J^μ is the conserved current corresponding to the symmetry. In the case of $\psi \to e^{i\theta}\psi$, the current is

---

The original version of this chapter was revised: The errors in this chapter have been corrected. The correction to this chapter can be found at https://doi.org/10.1007/978-3-030-56168-0_16

© The Author(s), under exclusive license to Springer Nature Switzerland AG 2020, corrected publication 2021
J. M. Cline, *Advanced Concepts in Quantum Field Theory*,
SpringerBriefs in Physics, https://doi.org/10.1007/978-3-030-56168-0_11

$$J^\mu = \bar{\psi}\gamma^\mu\psi. \tag{11.4}$$

If ψ was the electron field, then this would be simply the electric current (modulo the charge e). For a generic fermion field, it is the particle current, and its conservation implies that the net number of fermions of that kind does not change in any interaction. The current J^μ in (11.4) is called the *vector current* for the field ψ, since it transforms like a vector under the Lorentz group. This symmetry is sometimes denoted by $U(1)_V$.

We noted that for Majorana fermions, one cannot define a symmetry transformation $\psi \to e^{i\theta}\psi$; however, it does make sense to consider the axial or *chiral* transformation $\psi \to e^{i\theta\gamma_5}\psi$, which is denoted by $U(1)_A$. In this case we have

$$\begin{pmatrix} \sigma_2\psi_L^* \\ \psi_L \end{pmatrix} \to \begin{pmatrix} e^{i\theta}\sigma_2\psi_L^* \\ e^{-i\theta}\psi_L \end{pmatrix}, \tag{11.5}$$

which is self-consistent. Similarly, we could do the same transformation for Dirac fermions,

$$\psi = \begin{pmatrix} \psi_R \\ \psi_L \end{pmatrix} \to \psi = \begin{pmatrix} e^{i\theta}\psi_R \\ e^{-i\theta}\psi_L \end{pmatrix}. \tag{11.6}$$

In either case, the corresponding current is given by the *axial vector* current,

$$J_5^\mu = \bar{\psi}\gamma_5\gamma^\mu\psi. \tag{11.7}$$

The charge associated with this current, $Q_5 = \int d^3x\, J_5^0$, counts the number of right-handed fermions *minus* the number of left-handed ones. But in a massive theory, this is not a conserved quantity because the mass term $m(\psi_L^\dagger\psi_R + \psi_R^\dagger\psi_L)$ can convert left-handed states into right-handed ones and *vice versa*. For Majorana particles, the corresponding statement is that the mass term can flip a left-handed particle into a right-handed antiparticle, so the notion of a conserved particle number is spoiled. Only if the theory is massless is the axial rotation a symmetry. We can see this using the classical equation of motion:

$$i\partial_\mu J_5^\mu = \bar{\psi}\gamma_5(i\overleftarrow{\partial} + i\overrightarrow{\partial})\psi = \bar{\psi}\gamma_5(m+m)\psi. \tag{11.8}$$

Only if $m = 0$ is the axial symmetry a good one for the classical Lagrangian.

Now for the surprising subject of this section: even though the vector and axial-vector currents may be conserved in the classical theory, it may be impossible to preserve both of these symmetries once loop effects are introduced. In the present case, there is no regularization scheme that respects both symmetries. Dimensional regularization does not work because there is no generalization of γ_5 to higher dimensions. Pauli-Villars fails because the masses of the regulator fields do not respect chiral symmetry. We shall now show that in the case of a single Dirac fermion, it is possible to preserve one or the other of the two symmetries, but not both [7].

The quantity where the problem shows up is in the Green's functions of composite operators which are products of three currents. Define

$$-i\Gamma_{\mu\nu\lambda} = \langle 0|T^*(J_{5\mu}(x_3)J_\nu(x_1)J_\lambda(x_2)|0\rangle. \qquad (11.9)$$

On the basis of the classical symmetries, we would expect $\Gamma_{\mu\nu\lambda}$ to satisfy the three *Ward identities* for conservation of the currents:

$$\partial_{x_3^\mu}\Gamma^{\mu\nu\lambda} = \partial_{x_1^\nu}\Gamma^{\mu\nu\lambda} = \partial_{x_2^\lambda}\Gamma^{\mu\nu\lambda} = 0. \qquad (11.10)$$

We will show that, due to the contribution of the *triangle diagrams* in Fig. 11.1, it is impossible to satisfy all of the above relations.

There are two diagrams because of the two possible ways of contracting the fermion fields. In the figure we have gone to momentum space, and in the second diagram we have added an arbitrary shift a to the momentum in the internal line, which correponds to a change of integration variables $k \to k + a$ in the integral for the second diagram. If everything was well-behaved, the choice of a should have no effect on the final answer. However, we will find that because the diagrams are UV divergent, the choice of a *does* matter. The result of evaluating the diagrams is that the divergences in (11.10) are given by

$$p_3^\mu\Gamma_{\mu\nu\lambda} = -\frac{1}{8\pi^2}\epsilon_{\nu\lambda\alpha\beta}a^\alpha p_3^\beta \qquad (11.11)$$

$$p_1^\nu\Gamma_{\mu\nu\lambda} = -\frac{1}{8\pi^2}\epsilon_{\lambda\mu\alpha\beta}(a+2p_2)^\alpha p_1^\beta \qquad (11.12)$$

$$p_2^\lambda\Gamma_{\mu\nu\lambda} = -\frac{1}{8\pi^2}\epsilon_{\mu\nu\alpha\beta}(a-2p_1)^\alpha p_2^\beta. \qquad (11.13)$$

To obtain this result, we can use a fact from vector calculus which saves us from having to evaluate the loop integrals in great detail. Namely, if we cut off the loop integral by a momentum space cutoff in Euclidean space, then it is an integral over a finite volume, the 4-sphere. We can write

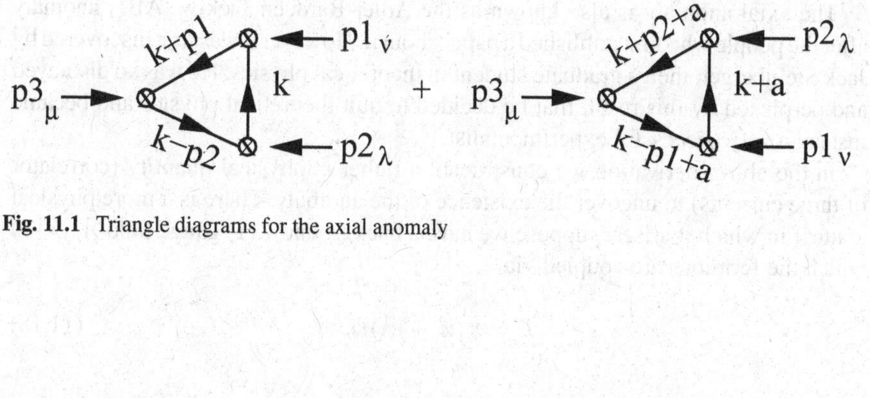

Fig. 11.1 Triangle diagrams for the axial anomaly

$$I(a) - I(0) = \int_{S_4} d^4k \, (f(k+a) - f(k))$$

$$= \int_{S_4} d^4k \, (a^\mu \partial_\mu f(k) + O(a^2))$$

$$= \mathbf{a} \cdot \int_{S_3} d\mathbf{S} \, f(k) + \cdots = a^\mu \int_{S_3} d\Omega_k k^2 k_\mu f |_{k^2 \to \infty} . \quad (11.14)$$

For an integral which is linearly divergent in k before subtracting the two terms, we can ignore the terms of higher order in a, because these involve more derivatives with respect to k, and their corrseponding surface terms vanish at infinity. For example, if $f = k_\nu / (k^2 + b^2)^2$, then

$$I(a) - I(0) = a^\mu \int d\Omega_k k^2 \frac{k_\mu k_\nu}{(k^2)^2}$$

$$= \frac{1}{4} a^\mu \delta_{\mu\nu} \int d\Omega_k k^2 \frac{k^2}{(k^2)^2} = \frac{\pi^2}{2} a^\nu . \quad (11.15)$$

The dependence on the arbitrary vector a can be considered to be part of our choice of regularization. For a general choice of a, neither the vector nor axial vector currents are conserved (although there might some linear combination of them which *is* conserved). We can make one choice which preserves the vector symmetry however, which is the relevant one if we are talking about electrons since we certainly want electric charge to be conserved:

$$a = 2(p_1 - p_2) . \quad (11.16)$$

With this choice, both of the vector current divergences (11.12) and (11.13) vanish, and the axial anomaly takes on a definite value,

$$-i p_3^\mu \Gamma_{\mu\nu\lambda} = \frac{1}{2\pi^2} \epsilon_{\nu\lambda\alpha\beta} p_1^\alpha p_2^\beta . \quad (11.17)$$

The axial anomaly is also known as the Adler-Bardeen-Jackiw (ABJ) anomaly after the people who first published a paper about it. However it was first discovered by Jack Steinberger, then a graduate student in theoretical physics. He was so dismayed and perplexed by this result that he decided to quit theoretical physics and became instead a very successful experimentalist.

In the above derivation we constructed a rather unphysical quantity (correlator of three currents) to uncover the existence of the anomaly. There is a more physical context in which it arises; suppose we have a background U(1) gauge field $A_\mu(x)$ to which the fermions are coupled via

$$\mathcal{L} = \bar{\psi}(i\partial\!\!\!/ + e A\!\!\!/)\psi . \quad (11.18)$$

With the gauge interaction, it is no longer necessary to consider the 3-current Green's function; we can compute $\langle \partial_\mu J_5^\mu \rangle$ directly. The other two vertices in the triangle diagram arise from perturbing to second order in the interaction. We get an extra factor of $\frac{1}{2!}$ from second order perturbation theory, so (11.17) becomes

$$p_{3\mu} J_5^\mu(p_3) = \frac{e^2}{4\pi^2} \epsilon_{\nu\lambda\alpha\beta} p_1^\alpha A^\nu(p_1) p_2^\beta A^\lambda(p_2)$$

$$= \frac{e^2}{16\pi^2} \epsilon_{\nu\lambda\alpha\beta} \left(p_1^\alpha A^\nu(p_1) - p_1^\nu A^\alpha(p_1) \right) \left(p_2^\beta A^\lambda(p_2) - p_2^\lambda A^\beta(p_2) \right) . \quad (11.19)$$

Transforming back to position space, this gives

$$\partial_\mu J_5^\mu = \frac{e^2}{16\pi^2} \epsilon_{\nu\lambda\alpha\beta} F^{\nu\lambda} F^{\alpha\beta} \equiv \frac{e^2}{16\pi^2} F^{\nu\lambda} \widetilde{F}_{\nu\lambda} . \quad (11.20)$$

The quantity $F^{\nu\lambda} \widetilde{F}_{\nu\lambda}$ can be shown to be the same as $2\mathbf{E} \cdot \mathbf{B}$. If we integrate (11.20) over space,

$$\int d^3x\, \partial_\mu J_5^\mu = \dot{Q}_5 = \frac{e^2}{8\pi^2} \int d^3x\, \mathbf{E} \cdot \mathbf{B} . \quad (11.21)$$

This has a remarkable interpretation: it says that if electrons were massless, then in the presence of a region of space with parallel electric and magnetic fields, axial charge would be created at a constant rate given by (11.21). In other words, there would be a constant rate of production of pairs of left-handed particles and right-handed antiparticles, or *vice versa*. These particles would start out with zero energy (remember that they are massless), but they would be accelerated by the electric field thereafter.

The way this happens can be visualized in terms of the migration of single particle states as shown in Fig. 11.2. If we put the system in a box, only discrete momentum values will be allowed. Due to the electric fields, the momentum of a given state will increase linearly with time. Let's focus on states with spin $S_z = +1/2$. We would call them left-handed if they are moving with negative group velocity ($p_z < 0$) and right-handed if $p_z > 0$—except for the negative energy particles in the filled Dirac sea we interpret things oppositely since these denote the absence of the corresponding antiparticle. When there is a mass gap m and the electric field is smaller than m^2, no transitions between negative and positive energy states take place, so the chirality violation is not observable (it just corresponds to a constant production of new LH states in the filled Dirac sea). If $m = 0$, the gap disappears, and then the filled states of the Dirac sea jump to occupy positive energy states regardless of how small the electric field is. The production of chiral pairs (particles with $S_z = +1/2$ and antiparticles with $S_z = -1/2$, moving in opposite directions) occurs and chirality violation is observable. But one might wonder what the role of the magnetic field is in all of this—these pictures make no reference to B after all. Indeed if $B = 0$, we could draw the same diagrams for the $S_z = -1/2$ states, and these would produce an equal and opposite amount of chiral charge to the $S_z = +1/2$ states. However, in the

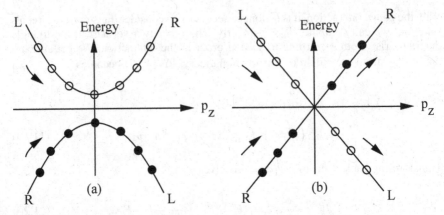

Fig. 11.2 Creation of axial charge in parallel electric and magnetic fields for **a** massive and **b** massless electrons

presence of the magnetic field, we have new contributions to the energy of a particle; there are Landau levels due to the electrons circulating around the B field lines, and there is also a magnetic moment coupling to the spin, so that the dispersion relation takes the form

$$E^2 = p_z^2 + (2n + 1)eB + -2eBS_z + m^2 . \qquad (11.22)$$

Therefore only the states with Landau level $n = 0$ and $S_z = +1/2$ have zero energy; the rest get an effective contribution to their mass. This heuristic discussion makes the physical origin of the anomaly more plausible.

There are numerous ways of deriving the axial anomaly, which give equivalent results. One way is to regularize the axial current by point-splitting:

$$J_5^\mu = \lim_{\epsilon_\mu \to 0} \bar{\psi}(x + \epsilon/2)\gamma^\mu\gamma_5 \exp\left(i \int_{x-\epsilon/2}^{x-\epsilon/2} dx^\nu A_\nu(x) \right) \psi(x + \epsilon/2) . \qquad (11.23)$$

The exponential factor is required if one wants to keep the current invariant under the electromagnetic gauge symmetry:

$$\psi \to e^{ie\Omega(x)}\psi; \qquad A_\mu \to A_\mu + \partial_\mu\Omega . \qquad (11.24)$$

The small separation vector ϵ_μ cuts off the UV divergence of the momentum integral in the triangle anomaly, and gives the result we have computed when one averages over the directions of ϵ_μ to restore Lorentz invariance.

Yet another derivation (due to Fujikawa [12]) derives the axial anomaly from the path integral, by showing that the path integral measure is not invariant under the $U(1)_A$ symmetry when one carefully defines it in such a way as to preserve the $U(1)_V$ symmetry. Consider a theory with several fields which we will collectively denote by ϕ, having action $S[\phi]$ and some symmetry transformation $\phi \to \phi + \epsilon A\phi$, where A is a matrix. In evaluating the path integral, we can change variables from ϕ to $\phi' = \phi + \epsilon A\phi$,

$$\int \mathcal{D}\phi e^{iS[\phi]} = \int \mathcal{D}\phi' e^{iS[\phi']}$$

$$= \int \mathcal{D}\phi' e^{iS[\phi]+i\int d^4x \partial_\mu \epsilon J^\mu}$$

$$= \int \mathcal{D}\phi' e^{iS[\phi]} \left(1 + i\int d^4x \partial_\mu \epsilon J^\mu \right). \tag{11.25}$$

If it is true that $\mathcal{D}\phi' = \mathcal{D}\phi$, then we see after integrating by parts that

$$\int \mathcal{D}\phi e^{iS[\phi]} \int d^4x \epsilon \partial_\mu J^\mu = 0$$

$$\longrightarrow \langle \partial_\mu J^\mu \rangle = 0. \tag{11.26}$$

The latter equality follows from the fact that $\epsilon(x)$ can be any function, including a delta function with support at an arbitrary position.

The above argument can be applied to the U(1)$_A$ symmetry, and it would imply that there is no anomaly if it was true. The one questionable assumption we made in the derivation was that $\mathcal{D}\phi' = \mathcal{D}\phi$. Evidently this is not true in the case of the U(1)$_A$ transformation. If we define the transformation as

$$\psi \to e^{i\epsilon\gamma_5/2}\psi, \quad \bar{\psi} \to \bar{\psi}e^{i\epsilon\gamma_5/2} \tag{11.27}$$

then the Nöther current has the conventional normalization, and the Jacobian is

$$\left| \frac{\partial(\psi', \bar{\psi}')}{\partial(\psi, \bar{\psi})} \right| = \left| \begin{matrix} e^{i\epsilon\gamma_5/2} & 0 \\ 0 & e^{i\epsilon\gamma_5/2} \end{matrix} \right| = e^{i\epsilon\gamma_5} \tag{11.28}$$

for the field at each point x. We must multiply these together for all x, so the Jacobian is $\det e^{i\epsilon\gamma_5}$ in field space. Let us compute

$$\ln \det e^{i\epsilon\gamma_5} = \text{tr} \ln(e^{i\epsilon\gamma_5}) = \text{tr} \ln(1 + i\epsilon\gamma_5) = i\,\text{tr}\,\epsilon\,\gamma_5. \tag{11.29}$$

Since trγ_5 vanishes naively, we might think that this can lead nowhere. However since this is a trace in field space, we have to find all the eigenvalues of the Dirac operator and sum the expectation value of γ_5 over them. If there are more left-handed than right-handed solutions, we can get a nonvanishing result. We know that the anomaly shows up most readily if we quantize in a background U(1)$_V$ gauge field, so let us do this. Then the explicit meaning of tr γ_5 is

$$\text{tr}\, \epsilon\, \gamma_5 = \sum_n \int d^4 x\, \epsilon\, \bar\psi_n \gamma_5 \psi_n, \tag{11.30}$$

where the functions are eigenfunctions of the Dirac operator in a background gauge field:

$$(i\partial\!\!\!/ + e A\!\!\!/)\psi_n = \lambda_n \psi_n . \tag{11.31}$$

Notice that we can rewrite the fields as $\psi(x) = \sum_n a_n \psi_n(x)$, $\bar\psi(x) = \sum_n b_n \bar\psi_n(x)$, where a_n and b_n are Grassmann numbers, and the path integral becomes

$$\int \mathcal{D}\psi \mathcal{D}\bar\psi e^{iS} = \prod \int da_n db_n \, \exp \sum_n b_n \lambda_n a_n = \prod \lambda_n = \det(i\partial\!\!\!/ + e A\!\!\!/) . \tag{11.32}$$

Now the problem is that the sum in (11.30) is not well-defined without being regulated in the ultraviolet, and we must do so in a way that respects the $U(1)_V$ symmetry. Doing so carefully results in a divergence for the axial current that agrees with our previous derivations [13].

Chapter 12
Abelian Gauge Theories: QED

We have focused on some toy model theories so far, to keep things relatively simple. Although ϕ^4 and Yukawa interactions are part of the standard model, they appear in a somewhat more complicated way than we have treated them. Gauge interactions are the major ingredient we have not yet discussed. In this section we will treat the Abelian case, which includes QED.

You were already exposed to the quantization of gauge fields in your previous course. Let's review. The gauge field can be derived from promoting the global U(1)$_V$ symmetry of a Dirac fermion $\psi \to e^{i e \theta}\psi$ to a local symmetry, $\psi \to e^{i e \theta(x)}\psi$. We have now factored out the charge of the electron e in our definition of the symmetry operation, since a particle with a different charge, like the up quark, would transform as $u \to e^{i \frac{2}{3} e \theta(x)} u$. The new term which arises due to the spatial variation of θ is

$$\delta \mathcal{L} = \bar{\psi}(-\not{\partial}\theta)\psi . \tag{12.1}$$

This can be canceled by adding the gauge interaction

$$\mathcal{L}_{\text{int}} = \bar{\psi}(e\not{A})\psi , \tag{12.2}$$

if we allow the gauge field to transform as

$$A_\mu \to A + \partial_\mu \theta , \tag{12.3}$$

which of course is just a gauge transformation, which leaves the field strength $F_{\mu\nu}$ unchanged. The action for the gauge field is

$$\mathcal{L}_{\text{gauge}} = -\frac{1}{4} F_{\mu\nu} F^{\mu\nu} = \frac{1}{2}\left(\vec{E}^2 - \vec{B}^2\right) . \tag{12.4}$$

To get more insight, let's make a choice of gauge, say Coulomb gauge, where $\vec{\nabla} \cdot \vec{A} = 0$. The Lagrangian density can be written as

---

The original version of this chapter was revised: The errors in this chapter have been corrected. The correction to this chapter can be found at https://doi.org/10.1007/978-3-030-56168-0_16

© The Author(s), under exclusive license to Springer Nature Switzerland AG 2020, corrected publication 2021
J. M. Cline, *Advanced Concepts in Quantum Field Theory*,
SpringerBriefs in Physics, https://doi.org/10.1007/978-3-030-56168-0_12

$$\frac{1}{2}\left((\dot{\vec{A}} - \vec{\nabla}A_0)^2 - (\vec{\nabla}\times\vec{A})^2\right) = \frac{1}{2}\left(\dot{\vec{A}}^2 + (\vec{\nabla}A_0)^2 - (\vec{\nabla}\times\vec{A})^2 - 2\vec{\nabla}A_0\cdot\dot{\vec{A}}\right)$$

$$\text{integrate by parts} \rightarrow \frac{1}{2}\left(-\vec{A}\partial_0^2\vec{A} + \vec{A}\cdot\vec{\nabla}^2\vec{A} - A_0\vec{\nabla}^2 A_0\right), \quad (12.5)$$

where we used the fact that $\vec{\nabla}\cdot\vec{A} = 0$ to get rid of the term $2\vec{\nabla}A_0\cdot\dot{\vec{A}}$. Roughly speaking, $\vec{E}$ looks like the time derivative and $\vec{B}$ looks like the spatial derivative of the three components of $\vec{A}$ (which therefore resemble three scalar fields), but we also get the zeroth component, which is strange because it has no kinetic term and its gradient term has the wrong sign. If we couple the gauge field to an external electromagnetic current J_μ, the path integral over the gauge fields is

$$\int \mathcal{D}A_\mu e^{iS[A]-i\int d^4x A^\mu J_\mu}. \quad (12.6)$$

We can rewrite the exponent as

$$\int d^4x\left(\frac{1}{2}\left[-\vec{A}\Box\vec{A} + A_0\vec{\nabla}^2 A_0\right] - A_0 J_0 + \vec{A}\cdot\vec{J}\right)$$

$$= \frac{1}{2}\left[-\left(\vec{A} - \frac{1}{\Box}\vec{J}\right)\Box\left(\vec{A} - \frac{1}{\Box}\vec{J}\right) + \left(A_0 + \frac{1}{\vec{\nabla}^2}J_0\right)\vec{\nabla}^2(A_0 + \frac{1}{\vec{\nabla}^2}J_0)\right]$$

$$+ \frac{1}{2}\left[\vec{J}\cdot\frac{1}{\Box}\vec{J} - J_0\frac{1}{\vec{\nabla}^2}J_0\right] \quad (12.7)$$

By shifting the integration variable in the functional integral, we see the result is

$$\int \mathcal{D}A_\mu e^{iS[A]-i\int d^4x A^\mu J_\mu} = e^{\frac{i}{2}\int d^4x\left[\vec{J}\cdot\frac{1}{\Box}\vec{J}-J_0\frac{1}{\vec{\nabla}^2}J_0\right]}\int \mathcal{D}A_\mu e^{iS[A]}$$

$$= e^{-\frac{i}{2}\int d^4x A_\mu[J]J^\mu} \quad (12.8)$$

where $A_\mu[J]$ is the gauge field which is induced by the external current in Coulomb gauge:

$$A_i[J] = \frac{1}{\Box}J_i; \qquad A_0[J] = \frac{1}{\vec{\nabla}^2}J_0. \quad (12.9)$$

The exponent in (12.8) is the action due to the electromagnetic interaction of the current with itself, so the result makes sense.

However, I was sloppy in my treatment of the path integral in this derivation. How is the constraint that $\vec{\nabla}\cdot\vec{A} = 0$ implemented in the path integral? You may have already learned about this—it is the Faddeev–Popov prodedure. The problem with trying to do the path integral (12.6) without gauge fixing is that it does not exist due to an infinite factor associated with the gauge symmetry. Namely, at each point in spacetime, one can change A_μ by the gauge transformation $A_\mu \rightarrow A_\mu + \partial_\mu \Lambda(x)$ without affecting the action. The problem can be seen when we try to compute the gauge field propagator. In momentum space the action looks like

$$\frac{1}{2} \int \frac{d^4 p}{(2\pi)^4} A_\mu(-p) \left(-p^2 \eta^{\mu\nu} + p^\mu p^\nu \right) A_\nu(p), \tag{12.10}$$

but the matrix $p^2 \eta^{\mu\nu} - p^\mu p^\nu$ is singular; it has an eigenvector p_μ with zero eigenvalue. Therefore we cannot invert it to find the propagator. Of course, this eigenvector is proportional to the Fourier transform of a pure gauge configuration,

$$A_\mu = 0 \rightarrow A_\mu = \partial_\mu \Lambda$$
$$A_\mu(p) = 0 \rightarrow A_\mu = p_\mu \Lambda(p), \tag{12.11}$$

which must have vanishing action since $E = B = 0$ in this case. We would somehow like to avoid integrating over such gauge transformations when doing the path integral.

If we fix the gauge, this puts one constraint on the four fields A_μ, which means that only three of them are independent. The path integral measure therefore can be expressed in the form

$$\mathcal{D}A^\mu = \mathcal{D}A^\mu_{\text{g.f.}} \mathcal{D}\Lambda, \tag{12.12}$$

where $A_{\text{g.f.}}$ is constrained to lie in a lower-dimensional surface in the space of fields (the gauge fixing surface in Fig. 12.1) and the gauge transformations generated by Λ move each gauge-fixed field configuration (points in the subspace) off this surface, along curves called gauge orbits, along which the field strength does not change. Since the action does not depend on Λ, $\int \mathcal{D}\Lambda$ is the infinite factor which we want to remove from the path integral. The FP procedure extracts this infinity in a well-defined way, by inserting a cleverly designed factor of unity into the path integral before gauge fixing:

$$1 = \int \mathcal{D}\Lambda \, \delta[f(A_\mu + \partial_\mu \Lambda)] \det \left(\frac{\delta f(A + \partial \Lambda)_x}{\delta \Lambda_y} \right)$$
$$\equiv \int \mathcal{D}\Lambda \, \delta[f(A_\mu + \partial_\mu \Lambda)] \Delta_{\text{FP}}[A_\mu; \Lambda]. \tag{12.13}$$

Fig. 12.1 Gauge orbits piercing a gauge fixing surface

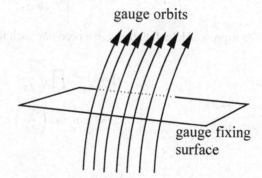

gauge orbits

gauge fixing surface

When we insert this into the path integral, we can use the fact that $S[A + \partial\Lambda] = S[A]$ to make the shift $A \to A - \partial\Lambda$ everywhere (since this is just shifting the integration variable), so that the path integral becomes

$$\int \mathcal{D}\Lambda\, \mathcal{D}A_\mu e^{iS[A]}\delta[f(A_\mu)]\Delta_{\text{FP}}[A_\mu; 0]. \tag{12.14}$$

The infinite factor of the gauge group volume now comes out cleanly since nothing in the remaining functional integral depends upon it.

The problem is now how to evaluate the remaining integral. Let's consider the Coulomb gauge choice again. In that case

$$f(A) = \vec{\nabla}\cdot\vec{A}, \tag{12.15}$$

and the FP Jacobian matrix is

$$\frac{\delta f(\partial\Lambda)_x}{\delta\Lambda_y} = \frac{\delta\left(\partial_\mu\eta^{\mu i}\partial_i\Lambda(x)\right)}{\delta\Lambda(y)} = \vec{\nabla}^2\delta(x-y). \tag{12.16}$$

The determinant can be expressed using

$$\ln\det\vec{\nabla}^2\delta(x-y) = \text{tr}\ln\vec{\nabla}^2\delta(x-y) = \int d^4x \int \frac{d^4p}{(2\pi)^4}\ln(\vec{p}^2), \tag{12.17}$$

similarly to our discussion of the vacuum diagrams $\bigcirc$. However, although this determinant is UV divergent, it does not depend on the gauge field. It is a harmless numerical factor which can be absorbed into the normalization of the path integral, and we can ignore it. More complicated choices of gauge would have resulted in a FP determinant which was a function of A_μ; this will often be unavoidable when we come to QCD in the next section.

We must still find a way of dealing with the delta functional which enforces the gauge condition. We can make things tractable by a suitable choice for how we represent the delta function. For example,

$$\delta(x) = \lim_{M\to\infty}\sqrt{\frac{M}{2\pi}}e^{-M^2x^2/2}. \tag{12.18}$$

To express a delta functional, we need one such factor for each point of spacetime:

$$\delta[f(x)] = \lim_{C\to\infty}\prod_{x^\mu}\sqrt{\frac{C}{2\pi}}e^{-C^2f^2(x)/2}$$

$$= \lim_{C\to\infty}\det\left(\frac{C}{2\pi}\right)e^{-\frac{1}{2}C^2\int d^4x\, f^2(x)}. \tag{12.19}$$

Like the FP determinant, $\det(C/2\pi)$ is another harmless infinity we can absorb into the measure. (Remember that the vacuum generating functional is always normalized to be unity anyway). Now in the case of $f = \vec{\nabla} \cdot \vec{A}$, it would be nice if the exponential term looked more like part of the action. We can accomplish this by taking $C^2 \to -iC^2$ (this is also a good representation of a delta function–the oscillations rather than exponential decay are what kill the integral away from the support of the delta function), so for Coulomb gauge we get

$$\delta[f(A)] = \lim_{C\to\infty} e^{i\frac{1}{2}C^2 \int d^4x\, (\vec{\nabla}\cdot\vec{A})^2}. \tag{12.20}$$

The action now becomes

$$\frac{1}{2}\int \frac{d^4p}{(2\pi)^4} A_\mu(-p)\left(-p^2\eta^{\mu\nu} + p^\mu p^\nu - C^2 p_\perp^\mu p_\perp^\nu\right) A_\nu(p), \tag{12.21}$$

where we defined the transverse momentum vector $p_\perp^\mu$ as just the spatial components:

$$p_\perp^\mu = p^\mu - \delta_0^\mu p^0. \tag{12.22}$$

One can check that for nonzero values of C, the tensor $-p^2\eta^{\mu\nu} + p^\mu p^\nu - C^2 p_\perp^\mu p_\perp^\nu$ is now nonsingular and therefore invertible—the propagator exists.

What about taking the limit $C \to \infty$? We now see that this is not really necessary. We can use any nonzero value of C and still define the propagator. In fact, this is generally more convenient than taking $C \to \infty$. For example when $C = 1$, most of the terms in $-p^\mu p^\nu + C^2 p_\perp^\mu p_\perp^\nu$ cancel each other, making the propagator take a simpler form. In terms of our picture (Fig. 12.1), it means we are no longer integrating over a thin gauge fixing surface, but rather a fuzzy region centered about the thin slice, which nevertheless gives a finite result for the path integral. In other words, we are summing over a number of different gauge fixing surfaces, with different weights. There is nothing wrong with this; we only care about making the path integral finite.

There is another way we could arrive at the same result, which is somewhat more commonly given in textbooks. Instead of representing the delta functional by an exponential factor, suppose we have a true delta functional, but let the gauge condition be

$$\partial_i A_i = \omega(x), \tag{12.23}$$

for an arbitrary function $\omega(x)$. Then functionally average over different ω's with a Gaussian weight:

$$\delta[\partial_i A_i - \omega(x)] \to \int \mathcal{D}\omega\, e^{\frac{iC^2}{2}\int d^4x\, \omega^2}. \tag{12.24}$$

Not only does it lead to the same result as above, but it makes even more evident the physical picture of restricting the path integral to a "fuzzy" gauge fixing surface.

In the previous example, even with $C = 1$ the propagator still has a rather complicated form which is not Lorentz invariant. A much more convenient choice is the set of covariant gauges where we take $f = \partial_\mu A_\mu$ so that the action can be written as

$$\frac{1}{2} \int \frac{d^4 p}{(2\pi)^4} A_\mu(-p) \left(-p^2 \eta^{\mu\nu} + p^\mu p^\nu - p^\mu p^\nu / \alpha\right) A_\nu(p) \qquad (12.25)$$

(we have replaced C^2 by $1/\alpha$). Now when $\alpha = 1$ (Feynman gauge) we get the simple result that the photon propagator is given by

$$D_{\mu\nu}(p) = \langle A_\mu(-p) A_\nu(p)\rangle = \eta_{\mu\nu} \frac{-i}{p^2}. \qquad (12.26)$$

Notice the minus sign $(-i)$ relative to the Feynman rule for scalar field propagators. It is there solely to counteract the minus sign in η_{ij} for the spatial components. We want the fields which correspond to the real photon degrees of freedom (the components of $\vec{A}$ which are transverse to $\vec{p}$) to have the correct sign for their kinetic term. The longitudinal component A_0 has the wrong sign for its kinetic term. In a generic theory, this would be disastrous since a wrong sign kinetic term usually implies a breakdown of unitarity in the S matrix. We will see that gauge invariance saves us from this in gauge theories.

For the case $\alpha \neq 1$, we can find the propagator using Lorentz invariance. The propagator must have the form

$$D_{\mu\nu} = A \eta_{\mu\nu} + B p_\mu p_\nu, \qquad (12.27)$$

since these are the only Lorentz invariant tensors available. Contracting this with the inverse propagator, we obtain

$$D_{\mu\nu}(D^{-1})^{\nu\rho} = A p^2 \delta_\mu^{\ \rho} + \left[(1/\alpha - 1)A + B p^2 + (1/\alpha - 1)B p^2\right] = \delta_\mu^{\ \rho}, \quad (12.28)$$

which implies

$$A = \frac{1}{p^2}; \qquad B = \frac{\alpha - 1}{p^4}. \qquad (12.29)$$

At first it looks worrisome that the arbitrary parameter α is appearing in the propagator, since we are free to choose any value we like. Again, we will see that gauge invariance saves us: no physical quantities can depend on gauge parameters like α. Greens's functions and amplitudes will depend on α, but not S-matrix elements. This can provide a useful check on complicated calculations, if one is willing to do them in a general covariant gauge with α not fixed to be the most convenient value.

The fact is that S-matrix elements, so long as they only involve the physical states (the transverse photon polarizations, and not the longitudinal or time-like ones), do not depend on the gauge parameter, as long as the electrons are on their mass shells (i.e., they obey the Dirac equation $(\not{p} - m)\psi = 0$). Before proving this statement in

Fig. 12.2 Electron
self-energy in QED

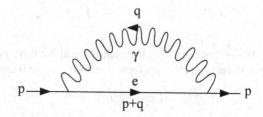

general, let's illustrate how it works for a one-loop diagram. The simplest example
which involves a photon propagator is the correction to the electron self-energy,
Fig. 12.2. In dimensional regularization, it is

$$i\Sigma(p) = (ie\mu^\epsilon)^2 \int \frac{d^4q}{(2\pi)^4} \gamma_\mu \frac{i}{\not{p}+\not{q}-m} \gamma_\nu \frac{-i}{q^2} \left(\eta_{\mu\nu} + \frac{\alpha-1}{q^2}q_\mu q_\nu\right). \quad (12.30)$$

Notice the factor of μ^ϵ which is needed to keep the charge dimensionless in $d = 4 - 2\epsilon$ dimensions. When we evaluate the divergent part of this diagram which is
proportional to the gauge-fixing term, we get

$$\Sigma_{\text{gauge-dep.}} \sim (1-\alpha)\frac{c}{\epsilon}(\not{p}-m). \quad (12.31)$$

This means that the mass counterterm will not depend on the gauge parameter. In the
renormalization procedure, even though one might first think that (12.31) requires
mass renormalization, we know that the true mass counterterm is determined after
doing wave function renormalization. The latter completely absorbs the α dependence in (12.31). Notice that whereas mass and charge are associated with physical
observables, wave function renormalization is not, so it is permissible to have α
dependence in the latter, though not in the former.

There are further restrictions on the form of the counterterms coming from gauge
invariance. These come from the Ward-Takahashi identities, which are relations
between Greens' functions which we can deduce in a similar manner to Fujikawa's
derivation of the axial anomaly–though now much simpler because in the present
case, the path integral measure really is invariant under gauge transformations, by
construction. The Ward identities are most easily stated in terms of the 1PI proper
vertices generated by the effective action. As we did for the axial anomaly, let us
first derive the procedure for a generic collection of fields denoted by ϕ_i, which
transforms linearly under an infinitesimal gauge transformation,

$$\delta\phi_i(x) = (A_{ij}\phi_j + A_i')\delta\omega(x), \quad (12.32)$$

where the inhomogeneous piece A_i' could be a differential operator (which is the
case for the change in the gauge field). Furthermore, suppose that the change in the
action is also linear in $\delta\phi_i(x)$,

$$\delta S = \int d^4x \, (B_i \phi_i + B') \delta \omega(x) . \tag{12.33}$$

We will assume that the path integral measure is invariant. Now consider how the path integral changes form when we change integration variables from ϕ to $\phi + \delta\phi$:

$$e^{iW[J]} = \int \mathcal{D}\phi \, e^{iS[\phi] + \int d^4x \, J_i \phi_i}$$

$$= \int \mathcal{D}\phi \, e^{iS[\phi] + \int d^4x \, J_i \phi_i} \left(1 + \int d^4x \, (B_i \phi_i + B_i' + J_i(A_{ij}\phi_j + A_i')) \, \delta\omega(x) \right) . \tag{12.34}$$

It follows that

$$0 = \int \mathcal{D}\phi \, e^{iS[\phi] + \int d^4x \, J_i \phi_i} \int d^4x \, (B_i \phi_i + B_i' + J_i(A_{ij}\phi_j + A_i')) \, \delta\omega(x) . \tag{12.35}$$

When we take the expectation value of ϕ in the presence of a classical source, this gives us precisely the classical field $\phi_c = \frac{\delta W}{\delta J}$. Furthermore $J = -\frac{\delta \Gamma}{\delta \phi_c}$, so we can rewrite (12.35) as

$$\int d^4x \, \frac{\delta \Gamma}{\delta \phi_c} \delta\phi_c = \int d^4x \, (B\phi_c + B') \delta\omega(x) = \int d^4x \, \delta S[\phi_c] . \tag{12.36}$$

But $\int d^4x \, \frac{\delta \Gamma}{\delta \phi_c} \delta\phi_c = \delta \Gamma[\phi_c]$, the change in the effective action. In other words, the change in the effective action is the same as the change in the original action. This is the general form of the Ward identities. We can now apply it to the specific case of QED.

For QED, we have

$$\delta A_\mu = \partial_\mu \delta\omega, \quad \delta\psi = ie\delta\omega\psi . \tag{12.37}$$

The action is invariant except for the gauge-fixing part,

$$S_{\text{g.f.}} = -\frac{1}{2\alpha} \int d^4x \, (\partial_\mu A^\mu)^2 \tag{12.38}$$

(note the minus sign corresponds to that in (12.25); we get one minus from $\partial^2 \to -p^2$ and another one from integrating by parts) so the change in the action is given by

$$\delta S = -\frac{1}{\alpha} \int d^4x \, \partial_\mu A^\mu \partial^2 \delta\omega . \tag{12.39}$$

This is the r.h.s. of the Ward identity. On the l.h.s we have, after integrating by parts,

$$\int d^4x \, \frac{\delta \Gamma}{\delta \phi_c} \delta\phi_c = e \int d^4x \, \delta\omega(x) \left(\frac{\delta \Gamma}{\delta \psi} i\psi + \frac{\delta \Gamma}{\delta \bar\psi}(-i\bar\psi) - \frac{1}{e} \partial_\mu \frac{\delta \Gamma}{\delta A_\mu} \right) . \tag{12.40}$$

The result is true for any function $\delta\omega$, in particular a delta function, so we can remove the integration and write

$$i\left(\frac{\delta\Gamma}{\delta\psi}\psi - \frac{\delta\Gamma}{\delta\bar\psi}\bar\psi\right) = \frac{1}{e}\left(\partial_\mu\frac{\delta\Gamma}{\delta A_\mu} - \frac{1}{\alpha}\partial^2\partial_\mu A^\mu\right). \tag{12.41}$$

All the fields here are understood to be the classical fields, although for clarity we have omitted the subscript. Equation (12.41) is in fact the starting point for an infinite number of relations between amplitudes because we can perform a functional Taylor expansion in the classical fields, and demand that the equation is true at each order. Let us first dispose of the gauge fixing term. At order A^1, ψ^0, $\bar\psi^0$, the only terms present are

$$\partial^\mu\Gamma^{(2A)}_{\mu\nu}A^\nu - \frac{1}{\alpha}\partial^2\partial_\nu A^\nu = 0$$
$$\longrightarrow \quad p^\mu\Gamma^{(2A)}_{\mu\nu}A^\nu + \frac{1}{\alpha}p^2 p_\nu A^\nu = 0. \tag{12.42}$$

At tree level, we know that the inverse propagator is

$$\Gamma^{(2A,\,\text{tree})}_{\mu\nu} = \left(-p^2\eta^{\mu\nu} + p^\mu p^\nu - p^\mu p^\nu/\alpha\right), \tag{12.43}$$

where the superscript $2A$ means we have two gauge fields as the external particles. Therefore the gauge-dependent term in the Ward identity exactly cancels its counterpart in the tree-level inverse propagator. Although the gauge-fixing term does get renormalized by loops, it remains true that it has no effect on the Ward identities involving external fermions or more than two gauge bosons.

An interesting consequence of the previous result is that beyond tree level, the vacuum polarization must be transverse:

$$p^\mu\Gamma^{(2A,\,\text{loops})}_{\mu\nu} = 0. \tag{12.44}$$

A more conventional notation instead of $\Gamma^{(2A,\,\text{loops})}_{\mu\nu}$ is $\Pi_{\mu\nu}$; we define it to be the part of the inverse photon propagator that comes from beyond tree level:

$$\Gamma^{(2A)} = \frac{1}{2}\int\frac{d^4p}{(2\pi)^4}A_\mu(-p)\left(-p^2\eta_{\mu\nu} + (1-\alpha^{-1})p_\mu p_\nu + \Pi_{\mu\nu}(p)\right)A_\nu(p). \tag{12.45}$$

Of course, Eq. (12.44) corresponds to our naive expectation that the electromagnetic current is conserved, since we can obtain $\Gamma^{(2A)}$ from the currents:

$$\Pi_{\mu\nu} = \langle J_\mu J_\nu\rangle = \langle(\bar\psi e\gamma_\mu\psi)(\bar\psi e\gamma_\nu\psi)\rangle. \tag{12.46}$$

The only way this can come about is if $\Pi_{\mu\nu}$ has the tensor structure

$$\Pi_{\mu\nu} \propto p^2 \eta_{\mu\nu} - p_\mu p_\nu \,, \tag{12.47}$$

since these are the only Lorentz-covariant tensors available. This is an example of how gauge invariance provides constraints that can give a nontrivial consistency check in practical calculations. The form (12.47) also has an important practical consequence: it gaurantees that the photon will not acquire a mass through loop effects. A mass term would have the form $\frac{1}{2}m_\gamma^2 A_\mu A^\mu$, which would show up as $m_\gamma^2 \eta_{\mu\nu}$ in the inverse propagator. The only way we could get this behavior would be if $\Pi_{\mu\nu}$ had the form

$$\Pi_{\mu\nu} \propto \left(p^2 \eta_{\mu\nu} - p_\mu p_\nu \right) \left(\frac{m_\gamma^2}{p^2} + \cdots \right). \tag{12.48}$$

This kind of behavior would imply that $\Pi_{\mu\nu}$ has an infrared divergence as $p^2 \to 0$ (notice that we can keep p_μ nonzero even though $p^2 = 0$ in Minkowski space). However, at least at one loop, one does not expect any infrared divergence because $\Pi_{\mu\nu}$ is generated by a fermion loop, and the internal electron is massive. In fact, even if it was massless, power counting indicates that the diagram would still be convergent in the infrared, since it has the form $\int d^4q/q^2$. On the other hand, in two dimensions, the integral is indeed IR divergent, and it is not hard to show that we indeed get a photon mass term as in (12.48), with the value $m_\gamma^2 = e^2/\pi$ (e has dimensions of mass in 2D), which was shown by Schwinger [13] to be an exact result. In fact, QED in 2D is known as the Schwinger model.

To avoid generating a photon mass in 4D, it is important to use a gauge invariant regularization. Notice that gauge invariance forbids us from writing an explicit mass term $\frac{1}{2}m_\gamma^2 A_\mu A^\mu$ in the Lagrangian. Therefore the massless of the photon is guaranteed if we do not spoil gauge invariance. This is different from gauge fixing. In the Faddeev-Popov procedure, we did not spoil gauge invariance in an arbitrary way, but carefully manipulated the path integral so that the underlying gauge symmetry is still present. If we arbitrarily change the Lagrangian by adding gauge noninvariant pieces, we can expect a photon mass to be generated by loops. This can be easily seen by examining the vacuum polarization diagram at zero external momentum:

$$
\begin{aligned}
i\Pi_{\mu\nu}(0) &= -(ie\mu^\epsilon)^2 \int \frac{d^d p}{(2\pi)^d} \mathrm{tr}\, \gamma_\mu \frac{\slashed{p}+m}{p^2-m^2} \gamma_\nu \frac{\slashed{p}+m}{p^2-m^2} \\
&= -(ie\mu^\epsilon)^2 d \int \frac{d^d p}{(2\pi)^d} \frac{(2p_\mu p_\nu - p^2 \eta_{\mu\nu} + m^2 \eta_{\mu\nu})}{(p^2-m^2)^2} \\
&= -(ie\mu^\epsilon)^2 d \, \eta_{\mu\nu} \int \frac{d^d p}{(2\pi)^d} \frac{(2/d - 1)p^2 + m^2}{(p^2-m^2)^2} \\
&= -(ie\mu^\epsilon)^2 d \, \eta_{\mu\nu} \int \frac{d^d p}{(2\pi)^d} \frac{(2/d - 1)(p^2 - m^2) + (2/d)m^2}{(p^2-m^2)^2}. \tag{12.49}
\end{aligned}
$$

If we were to regularize this using a momentum space cutoff instead of dimensional regularization, we would get a quadratically divergent contribution to the photon mass. But in DR, we get

$$i\Pi_{\mu\nu}(0) = -i(ie\mu^\epsilon)^2 \int \frac{d^d p_E}{(2\pi)^d} \left[\frac{2-d}{p_E^2 + m^2} - \frac{2m^2}{(p_E^2 - m^2)^2} \right]$$

$$= 2i \frac{(e\mu^\epsilon)^2}{(4\pi)^{2-\epsilon}} [(-1+\epsilon)\Gamma(-1+\epsilon) - \Gamma(\epsilon)] = 0 \,. \qquad (12.50)$$

Similarly, we could regularize using Pauli-Villars fields and also get zero, and this was the standard approach before the invention of DR. You can now appreciate (having done problem 11) what an improvement DR was.

The next interesting Ward identity comes from considering the terms of order A^0, ψ^1, $\bar\psi^1$:

$$i\left(\frac{\delta\Gamma^{(\bar\psi\psi)}}{\delta\psi(x)}\psi(x) - \frac{\delta\Gamma^{(\bar\psi\psi)}}{\delta\bar\psi(x)}\bar\psi(x) \right) = \frac{1}{e}\frac{\partial}{\partial x^\mu}\frac{\delta\Gamma^{(\bar\psi A\psi)}}{\delta A_\mu(x)} \,. \qquad (12.51)$$

Since $\Gamma^{(\bar\psi\psi)} = \int d^4x\, \bar\psi(i\slashed\partial - m)\psi$ and $\Gamma^{(\bar\psi A\psi)} = e\int d^4x\, \bar\psi\, \slashed A\psi$ at tree level, we obtain

$$i\int d^4y \left[\bar\psi(y)(i\slashed\partial_y - m)(-\delta(x-y))\psi(x) + \bar\psi(x)\delta(x-y)(i\slashed\partial_y - m)\psi(y) \right]$$

$$= \frac{1}{e}e \int d^4y\, \partial_{\mu,x}\bar\psi(y)\delta(x-y)\gamma^\mu\psi(y) \,, \qquad (12.52)$$

which indeed is an identity, since the mass terms cancel and the derivative terms match on either side of the equation. More generally, let $S(x, y)$ be the fermion propagator and $\Gamma_\mu(x, y, z)$ be the vertex function. Then we can write the Ward identity as

$$\frac{\partial}{\partial y_\mu}\Gamma_\mu(x, y, z) = -iS^{-1}(x-z)\delta(z-y), +iS^{-1}(x-z)\delta(x-y) \,, \qquad (12.53)$$

which in momentum space is

$$(p-k)_\mu\Gamma_\mu(p, p-k, k) = S^{-1}(p) - S^{-1}(k) \,. \qquad (12.54)$$

At tree level, $\Gamma_\mu = i\gamma_\mu$ and $S^{-1}(p) = i(\slashed p - m)$. This tells us nothing new yet.

However at the loop level, things become more interesting. To appreciate this, let's first look at the general structure of the renormalized Lagrangian for QED. Following the notation of F. Dyson,

$$\mathcal{L}_{ren} = Z_2\bar\psi i\slashed\partial\psi - mZ_m\bar\psi\psi + e\mu^\epsilon Z_1\bar\psi\slashed A\psi + \frac{1}{4}Z_3 F_{\mu\nu}F^{\mu\nu} + \frac{1}{2}Z_\alpha(\partial_\mu A^\mu)^2$$

$$= \bar\psi_0 i\slashed\partial\psi_0 - m_0\bar\psi_0\psi_0 + e_0\bar\psi_0\slashed A_0\psi_0 + \frac{1}{4}F_{0,\mu\nu}F_0^{\mu\nu} + \frac{1}{2\alpha_0}(\partial_\mu A_0^\mu)^2 \,, \qquad (12.55)$$

Fig. 12.3 Vacuum
polarization in QED

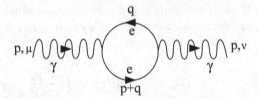

where we can see that the relation between bare and renormalized parameters is

$$\psi_0 = \sqrt{Z_2}\psi, \quad A_0^\mu = \sqrt{Z_3}A_0^\mu, \quad e_0 = e\mu^\epsilon \frac{Z_1}{Z_2\sqrt{Z_3}}, \quad m_0 = m\frac{Z_m}{Z_2}, \quad \alpha_0 = \frac{Z_3}{Z_\alpha}.$$
$$(12.56)$$

Were it not for gauge invariance, we would have to compute all three coefficients Z_1, Z_2, and Z_3, to obtain the beta function for the coupling e. But now at the loop level, the Ward identity is telling us that

$$Z_1(p-q)_\mu \Gamma_\mu(p, p-q, q) = Z_2(S^{-1}(p) - S^{-1}(q)), \qquad (12.57)$$

which implies that $Z_1 = Z_2$.

Therefore, we can compute the relation between the bare and renormalized charges without having to evaluate the more complicated diagram of Fig. 12.4, which determines Z_3. Instead, we can just compute the vacuum polarization diagram, Fig. 12.3. It is given by

$$i\Pi_{\mu\nu} = -(ie\mu^\epsilon)^2 \int \frac{d^{4+2\epsilon}q}{(2\pi)^{4+2\epsilon}} \text{tr}\left(\gamma_\mu \frac{i}{\not p + \not q - m}\gamma_\nu \frac{i}{\not q - m}\right). \qquad (12.58)$$

If we are only interested in Z_3, we can evaluate this in the limit of vanishing photon momentum. Let's define

$$\Pi_{\mu\nu} = \left(p^2\eta_{\mu\nu} - p_\mu p_\nu\right)\Pi(p^2). \qquad (12.59)$$

We see that the effective action generated by the loop diagram has the form

$$\frac{1}{4}(1 - \Pi(0))F_{\mu\nu}F^{\mu\nu}, \qquad (12.60)$$

so that we should start with the renormalized Lagrangian

$$\frac{1}{4}(1 - \Pi(0))^{-1}F_{0,\mu\nu}F_0^{\mu\nu} \qquad (12.61)$$

in order to keep the photon kinetic term correctly normalized. Hence we see that

$$Z_3 = (1 - \Pi(0))^{-1} \cong 1 + \Pi(0). \qquad (12.62)$$

Fig. 12.4 Vertex correction
in QED

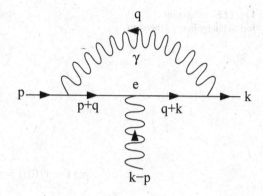

Now we can find the bare charge with respect to the renormalized charge, using
(12.56):

$$e_0 = \frac{e\mu^\epsilon}{\sqrt{Z_3}} \cong e\mu^\epsilon \left(1 - \frac{1}{2}\Pi(0)\right). \tag{12.63}$$

Computation of the divergent part of the vacuum polarization gives

$$\Pi(0) \sim -\frac{e^2}{12\pi^2\epsilon}. \tag{12.64}$$

The usual computation of the β function gives

$$\beta(e) = -\epsilon e + \frac{e^3}{12\pi^2}, \tag{12.65}$$

which shows that QED is not asymptotically free; the scale-dependent coupling runs
like

$$e^2(\mu) = \frac{e^2(\mu_0)}{1 - \frac{e^2(\mu_0)}{12\pi^2}\ln\frac{\mu^2}{\mu_0^2}}, \tag{12.66}$$

and the electric charge has a Landau singularity at high scales, like the quartic coupling did in $\lambda\phi^\mu$ theory.

The result (12.64) also tells us how the gauge-fixing term is renormalized. The
longitudinal part of the unrenormalized one-loop inverse propagator is

$$\begin{aligned}
p_\mu p_\nu (1 - \alpha^{-1}) + \Pi_{\mu\nu} &= (p^2\eta_{\mu\nu} \text{ part}) + p_\mu p_\nu (1 - \alpha^{-1} - \Pi(0)) \\
&\to p_\mu p_\nu (1 - \alpha^{-1} - \Pi(0))/(1 - \Pi(0)) \\
&= p_\mu p_\nu (1 - \alpha^{-1}(1 - \Pi(0))^{-1}).
\end{aligned}$$

The arrow above means "after renormalizing the gauge field wave function." We want
the divergent contribution to the effective action to be canceled by the counterterm
in $\frac{1}{2}Z_\alpha(\partial \cdot A)^2$, so we see that we should take

Fig. 12.5 Diagrams
forbidden by Furry's theorem

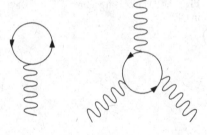

$$Z_\alpha = (1 - \Pi(0)) = 1/Z_3 \,. \tag{12.67}$$

We have touched upon the important one-loop effects in QED, Figs. 12.2, 12.3 and 12.4. But what about diagrams with an odd number of photons, as in Fig. 12.5? These would render the theory unrenormalizable if they were nonzero. It is easy to see that the first one vanishes, however:

$$\int \frac{d^4 p}{(2\pi)^4} \frac{\text{tr}[\gamma_\mu(\not{p} + m)]}{p^2 - m^2} = 0 \,. \tag{12.68}$$

Each term is zero either because of Dirac algebra or antisymmetry of the integrand. The second diagram is the triangle diagram in disguise, but without the γ_5. It is harder to directly show that it vanishes, but a simple observation shows that all diagrams with an odd number of external gauge bosons vanishes. This is Furry's theorem. The Lagrangian is invariant under charge conjugation symmetry:

$$\psi \to \psi^c = C\bar{\psi}^T, \qquad A_\mu \to -A_\mu \,. \tag{12.69}$$

This works because of the fact that $\bar{\psi}\gamma_\mu\psi = -\bar{\psi}^c\gamma_\mu\psi^c$, so the interaction term $\bar{\psi}\not{A}\psi$ remains invariant. Because of this symmetry of the original Lagrangian, the effective action must also have the symmetry. This includes diagrams with only external photons. For them, the amplitude must be invariant under simply $A_\mu \to -A_\mu$. This insures that if there is an odd number of external photons, the diagram vanishes. Intuitively, the reason the first diagram vanishes is that electrons and positrons give equal and opposite contributions to the electric potential. In a medium that has more electrons than positrons (unlike the vacuum), such diagrams would no longer be forced to vanish. Of course, this condition breaks the charge conjugation symmetry.

12.1 Applications of QED

A new technical problem arises in QED which we have not encountered before: infrared divergences. For some loop diagrams, the momentum integral can diverge for low momenta where virtual particles are going on their mass shell rather than

at high momenta. This problem typically arises when there is a massless particle
in the theory. The vacuum polarization diagram does not suffer from this problem
because the internal electrons only go on their mass shell when the external photon
momentum p goes to zero, but in this case we are saved by the explicit factors of
$p^2 \eta_{\mu\nu} - p_\mu p_\nu$ which must be in the numerator, thanks to gauge invariance. But the
diagrams like the electron self-energy and vertex correction do not have this saving
grace. The divergence can be seen for example in the vertex correction even before
doing any integrals,

$$
i\Gamma_\mu(p, p-k, k) = (ie\mu^\epsilon)^3 \int \frac{d^4 q}{(2\pi)^4} \gamma_\alpha \frac{i}{\slashed{q} + \slashed{k} - m} \gamma_\mu \frac{i}{\slashed{q} + \slashed{p} - m} \gamma_\beta \frac{-i\eta^{\alpha\beta}}{q^2}.
$$
(12.70)

Suppose that one of the electrons is on its mass shell, so $p^2 - m^2 = 0$. Then as
$q \to 0$, not only do we get a divergence from the photon propagator, but also the
electron propagator $(\slashed{p} - m)^{-1} = (\slashed{p} + m)/(p^2 - m^2)$. This would not happen if the
photon had a mass: it is kinematically imossible for an electron to decay into another
electron and a massive particle, whereas it is possible for an electron to emit a zero-
energy photon. This is why massless particles give rise to infrared divergences. In
the above example, assuming the external photon momentum is not also zero, the
second electron will have to be off shell, so we won't get an additional divergence
from the other propagator. One therefore expects a logarithmic IR divergence, and
this is what happens.

There are two ways to handle this divergence; one is to put in a small regulator
mass m_γ for the photon, so that its propagator becomes $-i\eta^{\alpha\beta}/(q^2 - m_\gamma^2)$. One might
worry about the fact that this spoils gauge invariance, but it turns out that for QED,
this does not lead to any disastrous consequences—it is possible to add a mass to a
U(1) gauge boson without destroying the unitarity or renormalizability of the theory,
even though gauge invariance is lost. (The same statement does not hold true for
nonAbelian theories.) Another way of regulating the IR divergences is through DR:
there will be additional poles in ϵ associated with them.

However, unlike UV divergences, there is a general theorem that IR divergences
do not require further modifications of the theory like counterterms or renormalization. Instead, one has to carefully define what the physical observables are. It is
guaranteed that IR divergences will cancel out of these physical quantities. In the case
of the vertex function, the cancellation is somewhat surprising at first—it involves
the amplitude for *emission of an additional photon*, that is, Brehmsstrahlung. In
Fig. 12.6 I have illustrated this for the case of Compton scattering, but I could just as
well have omitted the second electron and assumed the first one was scattering off
of an external electromagnetic field. Note that we can never add two diagrams that
have different particles in the initial or final states—that would be like adding apples
and oranges. However, when the emitted photon has a sufficiently small energy, it
will be unobservable in any real experiment. A real experiment will not be able to
distinguish an event with just an electron in the final state from one which has an
electron and a very soft photon, whose energy is below the resolution of the detector.
Therefore, referring to the diagrams in Fig. 12.6, the cross section we measure will
have the form

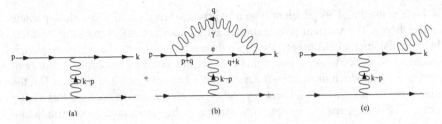

Fig. 12.6 Cancellation of IR divergences between the vertex correction and soft Brehmsstrahlung diagrams

$$\frac{d\sigma}{d\Omega}(e \to e) + \frac{d\sigma}{d\Omega}(e \to e\gamma) \sim |(a) + (b)|^2 + |(c)|^2$$

$$= |(a)|^2 + (a)(b)^* + (a)^*(b) + |(c)|^2 + O(e^8) . \quad (12.71)$$

The terms $(a)(b)^* + (a)^*(b) + |(c)|^2$ are all of the same order (e^6) in the perturbation expansion, so it makes sense to combine their effects. Amazingly, there is an IR divergence in (c) which exactly cancels the one coming from interference of the vertex correction with the tree diagram, $(a)(b)^* + (a)^*(b)$. This divergence is due to the fact that the virtual electron goes on shell (so its propagator blows up) as the photon energy goes to zero. The divergence comes from doing the phase space integral for the photon. It can also be cut off by introducing a small regulator mass for the photon.

The specific form of the IR-divergent parts of the above differential cross sections is (see p. 200 of [14])

$$\frac{d\sigma}{d\Omega}(e \to e) = \left(\frac{d\sigma}{d\Omega}(e \to e)\right)_0 \left[1 - \frac{\alpha}{\pi} \log\left(\frac{-l^2}{m^2}\right) \log\left(\frac{-l^2}{\mu^2}\right) + O(\alpha^2)\right] ;$$

$$\frac{d\sigma}{d\Omega}(e \to e\gamma) = \left(\frac{d\sigma}{d\Omega}(e \to e)\right)_0 \left[+\frac{\alpha}{\pi} \log\left(\frac{-l^2}{m^2}\right) \log\left(\frac{-l^2}{\mu^2}\right) + O(\alpha^2)\right] , \quad (12.72)$$

where $l = p - k$ is the momentum transfer carried by the t-channel photon, and these expressions are in the limit of large $-l^2$. Notice that the interference term is the negative one, which had to be the case since the cross section itself must be positive.

The Brehmsstrahlung cross section given above is integrated over all the emitted photons. To compare with a real experiment, we should only integrate over photons up to the experimental resolution E_r:

$$\frac{d\sigma}{d\Omega}(e \to e\gamma_{E<E_l}) = \left(\frac{d\sigma}{d\Omega}(e \to e)\right)_0 \left[+\frac{\alpha}{\pi} \log\left(\frac{-l^2}{m^2}\right) \log\left(\frac{E_l^2}{\mu^2}\right) + O(\alpha^2)\right] . \quad (12.73)$$

So the experimentally measured cross section has the form

$$\frac{d\sigma}{d\Omega}_{\text{expt.}} = \left(\frac{d\sigma}{d\Omega}(e \to e)\right)_0 \left[1 - \frac{\alpha}{\pi}\log\left(\frac{-l^2}{m^2}\right)\log\left(\frac{-l^2}{E_l^2}\right) + O(\alpha^2)\right], \quad (12.74)$$

still in the limit that $-l^2 \gg m^2$.

Before we leave the subject of QED, it should be said that the few computations we have touched upon don't begin to describe the fantastic successes of the theory in explaining experimental data. A notable example is the anomalous magnetic moment of the electron or muon. The electron couples to a magnetic field via

$$\mathcal{L}_{\text{int}} = -\vec{\mu} \cdot \vec{B} \qquad (12.75)$$

where the magnetic moment is related to the spin by

$$\vec{\mu} = g\left(\frac{e}{2m}\right)\vec{S} \qquad (12.76)$$

and g is the Landé g factor. At tree level, the Dirac equation predicts that $g = 2$. But this prediction is altered by the finite part of the vertex correction:

$$\frac{g-2}{2} = \frac{\alpha}{2\pi} = \frac{e^2}{8\pi^2} \qquad (12.77)$$

plus higher order corrections in α. These corrections are called the anomalous magnetic moment. In fact they have been computed to $O(\alpha^4)$. The calculations are so accurate that it is not useful to go to yet higher orders since the uncertainties due to loop diagrams with hadrons are greater than the size of the QED corrections. At the present, there is a small discrepancy between the predicted and measured values of $g - 2$ for the muon, which has led to a flurry of papers exploring how supersymmetric particles in the loops might be able to account for the result. The discrepancy is so far only at the 3σ level, however. To derive this result, it is useful to invoke the Gordon identity for the vector current of on-shell Dirac fermions:

$$\bar{u}(p')\gamma^\mu u(p) = \bar{u}(p')\left[\frac{p'^\mu + p^\mu}{2m} + \frac{i\sigma^{\mu\nu}(p'_\nu - p_\nu)}{2m}\right], \qquad (12.78)$$

where $-i\sigma^{\mu\nu} = \frac{1}{2}[\gamma^\mu\gamma^\nu]$. It is the $\sigma^{\mu\nu}$ part of the current which couples to the magnetic field, and one can see explicitly where the factor of e/m in the electron magnetic moment is coming from. Equation (12.78) is the tree-level current, which gets corrected at the loop level to the form

$$\Gamma^\mu(p', p) = \gamma^\mu F_1(q^2) + \frac{i\sigma^{\mu\nu}q_\nu}{2m}F_2(q^2), \qquad (12.79)$$

where $q = p' - p$ is the photon momentum, and the functions $F_i(q^2)$ are called form factors. These of course are obtained from a complete computation of the vertex function, including the finite parts. The magnetic moment interaction is defined in terms of a uniform magnetic field, whose momentum q is therefore zero, and so by comparing the above expressions we can deduce that

$$\frac{g-2}{2} = F_2(0). \tag{12.80}$$

One other very famous prediction is the Lamb shift, which is the splitting between the $2S_{1/2}$ and $2P_{1/2}$ states of the hydrogen atom. In the Dirac theory, these states are exactly degenerate because the spin-orbit interaction energy of the $2P_{1/2}$ state (using the tree-level value of the Landé g-factor) is exactly canceled by the relativistic correction to its kinetic energy, relative to that of the $2S_{1/2}$ state. In 1947, Willis Lamb and R. Retherford observed that the energy of the $2P_{1/2}$ state was lower than that of the $2S_{1/2}$ by an amount corresponding to the energy of a photon with a frequency of $\delta\omega = 1057.77$ MHz (or Mc/s). This was the experimental puzzle which motivated much of the early work on QED. The theoretical contributions which account for the Lamb shift consist of several parts: those from the two form factors F_i, and one from the vacuum polarization.

The contribution from F_2 is easy to understand; this just modifies the usual contribution of the splitting from the spin-orbit coupling in the obvious way. This is a small contribution to the total Lamb shift: only 68 MHz. The contribution from the vacuum polarization is also small, -27 MHz, but it has an interesting interpretation, which explains its mysterious name. The one-loop computation of $\Pi_{\mu\nu}$ in DR gives

$$\Pi(p) = \frac{e^2}{2\pi^2}\left[\frac{1}{6\epsilon} - \frac{1}{6\gamma} - \int_0^1 dx\, x(1-x)\ln\left[\frac{m^2 - p^2 x(1-x)}{\mu^2}\right]\right]. \tag{12.81}$$

To interpret it physically, we have to fix the finite part of the counterterms. We saw that by choosing $Z_3 = (1 - \Pi(0))^{-1}$, the photon wave function had the proper normalization at one loop. This tells us that the physically relevant part of $\Pi(p)$ is simply $\hat{\Pi}(p) = \Pi(p) - \Pi(0)$:

$$\hat{\Pi}(p) = -\frac{e^2}{2\pi^2}\int_0^1 dx\, x(1-x)\ln\left[\frac{m^2 - p^2 x(1-x)}{m^2}\right]. \tag{12.82}$$

Notice that this is the same integral which appeared in the one-loop correction for the quartic coupling in $\lambda\phi^4$ theory. We will come back to discuss its significance for large values of $p^2 > 4m^2$ later. For now, we are interested in $p^2 \ll m^2$ because the typical momenta of virtual photons in the hydrogen atom are of order the electron momentum, which is suppressed relative to m by a factor of α. Furthermore the Coulomb potential is static, so its Fourier transform depends only the spatial components of the momenta, which means we can take $p^2 \cong -\vec{p}^{\,2}$. In that limit,

$$\hat{\Pi}(p) = -\frac{e^2}{2\pi^2} \int\limits_0^1 dx\, x(1-x) \frac{\vec{p}^2 x(1-x)}{m^2} = -\frac{e^2}{60\pi^2} \frac{\vec{p}^2}{m^2} \qquad (12.83)$$

We would like to see how this result modifies the Coulomb potential. Recall our results (12.8) and (12.9). In the static limit, we can ignore the vector potential relative to the scalar one; it is the latter which gives the Coulomb self-interaction energy:

$$
\begin{aligned}
-S_{\text{coul}} = E_{\text{coul}} &= \frac{1}{2} \int d^3x \int d^3y\, \rho(x)\rho(y) \int \frac{d^3q}{(2\pi)^3} e^{i\vec{q}\cdot\vec{x}} \frac{-1}{|\vec{q}|^2[1-\hat{\Pi}(-|\vec{q}|^2)]} \\
&\cong -\frac{1}{2} \int d^3x \int d^3y\, \rho(x)\rho(y) \int \frac{d^3q}{(2\pi)^3} e^{i\vec{q}\cdot\vec{x}} \frac{1}{|\vec{q}|^2} \left[1 + \hat{\Pi}(-|\vec{q}|^2)\right] \\
&= +\frac{1}{2} \int d^3x \int d^3y\, \rho(x)\rho(y) V(|\vec{x}-\vec{y}|),
\end{aligned}
\qquad (12.84)
$$

where the modified Coulomb potential can be read off as

$$V(|\vec{x}|) = -\frac{1}{4\pi}\left(\frac{1}{|\vec{x}|} + \frac{4\pi e^2}{60\pi^2}\delta^{(3)}(\vec{x})\right). \qquad (12.85)$$

We see that the delta function contribution will lower the energy of the S-states of the hydrogen atom, but have no effect on the P states since their wave functions vanish at the origin. Hence it gives a negative contribution to the energy of $2S_{1/2}$ state.

The extra term was evaluated in the limit $|\vec{q}|^2 \ll m^2$. A more careful evaluation can be done. If we restore the factor of e^2 which is conventional to put in the Coulomb potential, the result is the Uehling potential, which is still an approximation that assumes $r \gg 1/m$:

$$V(|\vec{x}|) = -\frac{\alpha}{r}\left(1 + \frac{\alpha}{4\sqrt{\pi}} \frac{e^{-2mr}}{(mr)^{3/2}}\right) \equiv -\frac{\alpha(r)}{r}. \qquad (12.86)$$

In the limit $m \to \infty$, we recover the delta function, showing that the latter is a good approximation for distance scales much greater than the Compton wavelength of the electron. We have defined an effective distance-dependent coupling $\alpha(r)$ based on this result. We see that $\alpha(r)$ gets larger as one goes to shorter distances. We already knew this from another perspective: our calculation of the β function gave us the analogous result for α as a function of the energy scale, though it was valid for large energies, hence small distances. The two results give the same qualitative behavior however: the effective electric charge increases at short distances. This is where the picture of vacuum polarization comes in. We can understand this behavior if the vacuum itself is polarized by the presence of a bare electric charge, due to the appearance of virtual electron-positron pairs. These pairs will behave like dipoles which tend to screen the bare charge. At large distances, the screening is most effective, but as we get close to the bare charge, the effect of the screening is reduced.

So far we have accounted for only $68 - 27 = 41$ MHz of the total Lamb shift of 1058 MHz. The biggest contribution is coming from the F_1 form factor in the vertex correction. It is the most difficult part to compute, because it contains an infrared divergence. In the limit of small q^2, it can be shown that

$$F_1(q^2) = 1 + \frac{\alpha}{3\pi}\frac{q^2}{m^2}\left[\ln\left(\frac{m_\gamma}{m}\right) + \frac{3}{8}\right]. \tag{12.87}$$

(Sometimes you will see in addition to the term $\frac{3}{8}$ another term $\frac{1}{5}$, which is due to the vacuum polarization. We could push the latter into the vertex by doing a momentum-dependent field redefinition of the photon, but we have chosen not to do so.) To arrive at this result, we have chosen the finite parts of the counterterms in such a way that $F_1(0) = 1$; then the renormalized charge e couples to the electric current with the conventional normalization. Notice that q is the momentum of the photon. If we consider how F_1 corrects the Dirac equation, we get

$$(i\partial\!\!\!/ - m + eF_1(-\Box)A\!\!\!/)\psi = 0 \tag{12.88}$$

To treat the hydrogen atom, we would normally use $eA_\mu = \eta_{0\mu}\alpha/r$, the Coulomb potential. The above equation is telling us that we should correct this, effectively replacing the Coulomb potential by

$$\frac{\alpha}{r} \to \left(1 + F_1(\vec{\nabla}^2)\right)\frac{\alpha}{r}$$

$$= \frac{\alpha}{r} - \frac{4\alpha^2}{3m^2}\left[\ln\left(\frac{m_\gamma}{m}\right) + \frac{3}{8}\right]\delta^{(3)}(\vec{r}). \tag{12.89}$$

We can treat the extra term as a perturbation in the Schrödinger equation and compute its effect immediately for a given wave function. Like the vacuum polarization contribution, it has no effect on the P states. The only problem is that it is IR divergent; how do we interpret the photon mass?

To understand what is going on, one has to go back to the original form of the self-energy, a complicated expression, but the salient feature is that its IR divergence is only there because of the external electrons being on shell. In a bound state however, they are not exactly on shell; rather,

$$E = \sqrt{\vec{p}^2 + m^2} - E_b, \tag{12.90}$$

where $E_b \sim \alpha^2 mc^2$ is the binding energy. This means that

$$p^2 - m^2 \cong 2mE_b \tag{12.91}$$

and we should expect to replace m_γ^2 by something of order $|2mE_b|$. This crude procedure does in fact give a rather good estimate of the bulk of the Lamb shift. To do better and compute it exactly requires special techniques of bound state computations, and many pages of calculation.

Chapter 13
Nonabelian Gauge Theories

13.1 Group Theory for SU(N)

We have dealt with the simplest kind of gauge theory, which is based on the gauge symmetry $U(1)$. It is an Abelian group because any two $U(1)$ transformations commute with each other:

$$e^{i\theta_1} e^{i\theta_2} = e^{i\theta_2} e^{i\theta_1}. \tag{13.1}$$

On the other hand, $U(N)$, the group of $N \times N$ unitary matrices, is noncommutative: in general

$$U_1 U_2 \neq U_2 U_1 \tag{13.2}$$

for two such matrices. To express this in a way that looks more like (13.1), we use the fact that any $U(N)$ matrix can be written in the form

$$U = e^{i \sum_a \theta_a T^a} \tag{13.3}$$

where the matrices T_a are Hermitian, and are known as the *generators* of the group. The noncommutativity of the U(N) matrices in (13.2) is due to that of the generators. For two matrices which are only infinitesimally different from the unit matrix, $U_i \cong 1 + i\theta_{i,a} T^a$,

$$U_1 U_2 - U_2 U_1 = i\theta_{1,a}\theta_{2,b}[T^a, T^b]. \tag{13.4}$$

The generators obey the *Lie algebra*

$$[T^a, T^b] = if^{abc} T^c, \tag{13.5}$$

The original version of this chapter was revised: The errors in this chapter have been corrected. The correction to this chapter can be found at https://doi.org/10.1007/978-3-030-56168-0_16

© The Author(s), under exclusive license to Springer Nature Switzerland AG 2020, corrected publication 2021
J. M. Cline, *Advanced Concepts in Quantum Field Theory*,
SpringerBriefs in Physics, https://doi.org/10.1007/978-3-030-56168-0_13

where the *structure constants* f^{abc} are real and antisymmetric on any two indices. The number of different generators is called the *rank* of the group, and is given by N^2 for the group U(N).

However, the group U(N) is not *simple* except for the case of U(1): we can decompose any U(N) matrix into the product of a special unitary SU(N) matrix and a U(1) phase: U(N) = SU(N) × U(1). Therefore the rank of SU(N) is $N^2 - 1$. You are already familiar with SU(2), which is generated by the Pauli matrices:

$$T^a = \frac{1}{2}\sigma^a, \quad a = 1, 2, 3. \tag{13.6}$$

The factor of $\frac{1}{2}$ assures the generators are normalized in the conventional way

$$\text{tr}\, T^a T^b = \frac{1}{2}\delta_{ab}. \tag{13.7}$$

For SU(3), the generators are given by $T^a = \frac{1}{2}\lambda^a$, where the λ^a's are the Gell-Mann matrices,

$$\lambda_1 = \begin{pmatrix} 0 & 1 & 0 \\ 1 & 0 & 0 \\ 0 & 0 & 0 \end{pmatrix} \quad \lambda_2 = \begin{pmatrix} 0 & -i & 0 \\ i & 0 & 0 \\ 0 & 0 & 0 \end{pmatrix} \quad \lambda_3 = \begin{pmatrix} 1 & 0 & 0 \\ 0 & -1 & 0 \\ 0 & 0 & 0 \end{pmatrix}$$

$$\lambda_4 = \begin{pmatrix} 0 & 0 & 1 \\ 0 & 0 & 0 \\ 1 & 0 & 0 \end{pmatrix} \quad \lambda_5 = \begin{pmatrix} 0 & 0 & -i \\ 0 & 0 & 0 \\ i & 0 & 0 \end{pmatrix} \quad \lambda_6 = \begin{pmatrix} 0 & 0 & 0 \\ 0 & 0 & 1 \\ 0 & 1 & 0 \end{pmatrix}$$

$$\lambda_7 = \begin{pmatrix} 0 & 0 & 0 \\ 0 & 0 & -i \\ 0 & i & 0 \end{pmatrix} \quad \lambda_8 = \frac{1}{\sqrt{3}}\begin{pmatrix} 1 & 0 & 0 \\ 0 & 1 & 0 \\ 0 & 0 & -2 \end{pmatrix}.$$

$$\tag{13.8}$$

The structure constants for SU(3) are given by

abc	f_{abc}
123	1
147	1/2
156	−1/2
246	1/2
257	1/2
345	1/2
367	−1/2
458	$\sqrt{3}/2$
678	$\sqrt{3}/2$

$$\tag{13.9}$$

In addition, one can consider the anticommutators,

$$\{\lambda_a, \lambda_b\} = \frac{4}{3}\delta_{ab}I + 2d_{abc}\lambda_c\,, \tag{13.10}$$

where I is the $N \times N$ unit matrix. See the particle data book for a table of the d_{abc}'s for SU(3). Some other identities:

$$T^a T^b = \frac{1}{2N}\delta_{ab}I + \frac{1}{2}d_{abc}T^c + \frac{i}{2}f_{abc}T^c;$$

$$\sum_a T^a T^a = \frac{N^2-1}{2N}I \tag{13.11}$$

$$\sum_a T_{ij}^a T_{kl}^a = \frac{1}{2}\left(\delta_{il}\delta_{jk} - \frac{1}{N}\delta_{ij}\delta_{kl}\right)$$

$$\mathrm{tr}(T^a T^b T^c) = \frac{1}{4}(d_{abc} + if_{abc})$$

$$\mathrm{tr}(T^a T^b T^c T^d) = \frac{1}{4N}\delta_{ab}\delta_{cd} + \frac{1}{8}(d_{abe} + if_{abe})(d_{cde} + if_{cde})$$

$$f_{acd}f_{bcd} = N\delta_{ab} \tag{13.12}$$

13.2 Yang-Mills Lagrangian with Fermions

Now we would like to use this to do some physics. A fermion ψ_i which transforms in the fundamental representation of SU(N) will have an index i, which in the case of QCD (SU(3)) we call color, and in the case of the electroweak theory (SU(2)) we call isospin. It transforms as

$$\psi_i \rightarrow U_{ij}\psi_j = \left(e^{ig\sum_a \omega_a T^a}\right)_{ij}\psi_j \tag{13.13}$$

Here we have introduced a charge g which is analogous to the electric charge. To gauge this symmetry, we need to find a generalization of the covariant derivative. The ordinary derivative transforms as

$$\partial_\mu \psi \rightarrow U\partial_\mu \psi + (\partial_\mu U)\psi = U\left(\partial_\mu + U^{-1}(\partial_\mu U)\right)\psi\,. \tag{13.14}$$

The obvious way to cancel the unwanted term $(\partial_\mu U)\psi$ is to introduce the covariant derivative

$$D_\mu \psi = (\partial_\mu - igA_\mu^a T^a)\psi \tag{13.15}$$

In other words, the gauge field has become a Hermitian matrix which can be expanded in the basis T^a. It must transform like

$$A_\mu \equiv A_\mu^a T^a \rightarrow UA_\mu U^{-1} + \frac{i}{g}U\partial_\mu U^{-1} = UA_\mu U^\dagger + \frac{i}{g}U\partial_\mu U^\dagger\,. \tag{13.16}$$

Notice that this reduces to the usual expression in the Abelian case, where $U = e^{ig\Lambda(x)}$ for a simple function Λ: $A_\mu \to A_\mu + \partial_\mu \Lambda$.

The Feynman rules for the fermion propagator and the interaction of the gauge field with fermions are easy to read off from the interaction vertex. They are just like those of QED with the obvious changes,

$$\text{i} \longrightarrow \text{j} = i\frac{\delta_{ij}}{\not{p} - m};$$

$$\begin{array}{c} a,\mu \\ \end{array}$$

$$= ig\gamma_\mu T^a_{ij}$$

$$\text{i} \longleftarrow \text{j}$$

$$(13.17)$$

Next we need to specify the dynamics of the gauge field. In QED this was given by the Lagrangian $\frac{1}{4}F_{\mu\nu}F^{\mu\nu}$. For SU(N) we expect that the field strength will be a matrix. If we try to define it in the same way as in the U(1) case, $F_{\mu\nu} = \partial_\mu A_\nu - \partial_\nu A_\mu$, then under a gauge transformation we would get

$$F_{\mu\nu} \overset{?}{\to} UF_{\mu\nu}U^\dagger + \frac{i}{g}\partial_\mu\left(U\partial_\nu U^\dagger\right) - \frac{i}{g}\partial_\nu\left(U\partial_\mu U^\dagger\right)$$

$$\overset{?}{=} UF_{\mu\nu}U^\dagger + \frac{i}{g}[\partial_\mu U\partial_\nu U^\dagger - \partial_\nu U\partial_\mu U^\dagger]. \qquad (13.18)$$

If it weren't for the term in brackets, we could take the gauge field Lagrangian to be

$$\mathcal{L}_{\text{gauge}} = -\frac{1}{2}\text{tr}\,F_{\mu\nu}F^{\mu\nu} = -\frac{1}{4}\sum_a F^a_{\mu\nu}F^{a,\mu\nu}. \qquad (13.19)$$

But this does not work with F as we naively tried to define it. To discover the right definition for the field strength, we can appeal to a different identity which also works in the case of QED:

$$F_{\mu\nu} = \frac{i}{g}[D_\mu, D_\nu] = \partial_\mu A_\nu - \partial_\nu A_\mu - ig[A_\mu, A_\nu]. \qquad (13.20)$$

The new commutator term vanishes in the U(1) case, but under U(N) it transforms in just the right way to cancel the unwanted terms in (13.18). This term is the key to all that makes nonAbelian gauge theories very different from QED; without it, our theory would be equivalent to one with $N^2 - 1$ kinds of photons, i.e. a U(1)$^{N^2-1}$ gauge theory. When we expand (13.19), not only do we get the inverse propagator for the gauge fields, but also a cubic and a quartic interaction, as shown in Fig. 13.1.

We have already studied theories with cubic and quartic interactions, with no sign of dramatic effects like the confinement of quarks. Yet in the case of nonAbelian gauge theories, these new terms completely change the character of the physics, giving rise to asymptotic freedom at high energies, and confinement of quarks. Although

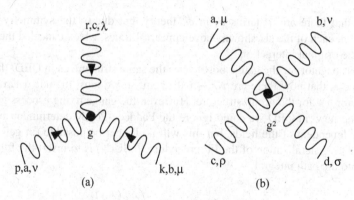

Fig. 13.1 New self-interactions of the gauge bosons in Yang-Mills theory

confinement cannot be seen in perturbation theory, we can at least show that the gauge coupling runs to larger values in the infrared, which is circumstantial evidence for confinement, since the interaction becomes increasingly strong. To study this, we need to find the Feynman rules for the new interactions. In position space, the interaction terms are

$$\mathcal{L}_{\text{int}} = -\frac{g}{2} f_{abc} A_b^\mu A_c^\nu (\partial_\mu A_\nu^a - \partial_\nu A_\mu^a) - \frac{g^2}{4} f_{abc} f_{ade} A_b^\mu A_c^\nu A_\mu^d A_\nu^e . \tag{13.21}$$

The first term can be simplified to $-g f_{abc} A_b^\mu A_c^\nu \partial_\mu A_\nu^a$ since the factor $f_{abc} A_b^\mu A^\nu$ is already antisymmetric under $\mu \leftrightarrow \nu$. In momentum space, we can write the corresponding action for this term as

$$S_3 = -g \int \frac{d^4 p}{(2\pi)^4} \frac{d^4 k}{(2\pi)^4} \frac{d^4 r}{(2\pi)^4} (2\pi)^4 \delta^{(4)}(p+k+r) f_{abc} i p_\mu A_\nu^a(p) A_b^\mu(k) A_c^\nu(r) . \tag{13.22}$$

To get the Feynman rule, we need to do all 6 possible contractions of the fields in (13.22) with the external fields whose momenta, color and Lorentz indices are shown in Fig. 13.1a. This is the same thing as taking functional derivatives, remembering that all the momenta and indices in (13.22) are dummy variables:

$$\begin{aligned}
\text{cubic Feynman rule} &= \frac{1}{(2\pi)^4 \delta^{(4)}(p+k+r)} i \frac{\delta^3 S_3}{\delta A_\nu^a(p) \delta A_\mu^b(k) \delta A_\lambda^c(r)} \\
&= g f_{abc} \eta_{\nu\lambda} p_\mu + \text{permutations of } (a\nu p, b\mu k, c\lambda r) \\
&= g f_{abc} \left[\eta_{\nu\lambda}(p-r)_\mu + \eta_{\nu\mu}(k-p)_\lambda + \eta_{\mu\lambda}(r-k)_\nu \right] . \tag{13.23}
\end{aligned}$$

Carrying out a similar procedure for the quartic coupling, we get

$$\begin{aligned}
\text{quartic Feynman rule} &= -ig^2 \Big[f^{abc} f^{cde}(\eta_{\mu\rho}\eta_{\nu\sigma} - \eta_{\mu\sigma}\eta_{\nu\rho}) + f^{ace} f^{bde}(\eta_{\mu\nu}\eta_{\rho\sigma} - \eta_{\mu\sigma}\eta_{\nu\rho}) \\
&\quad + f^{ade} f^{bce}(\eta_{\mu\nu}\eta_{\rho\sigma} - \eta_{\mu\rho}\eta_{\nu\sigma}) \Big] . \tag{13.24}
\end{aligned}$$

Notice that there are 4! terms as in ϕ^4 theory, but due to the symmetry of the interaction, each of the six shown above appears 4 times, which canceled the factor of $\frac{1}{4}$ in the original interaction.

The propagator for the gauge bosons has the same structure as in QED; the only difference is that now there are $N^2 - 1$ of them, so we must include a Kronecker delta factor δ^{ab} for the group structure. However, the gauge fixing process gives us something new; in QED we could ignore the Faddeev-Popov determinant because it was independent of the fields, but this will no longer be possible (in general) in SU(N). The generalization of the FP procedure to SU(N) is to insert the following factor into the path integral:

$$1 = \int \mathcal{D}U \, \delta[f(A_U)] \det \left(\frac{\delta f(A_U)_x}{\delta U_y} \right)$$

$$\equiv \int \mathcal{D}U \delta[f(A_U)] \Delta_{\text{FP}}[A; U] \,, \tag{13.25}$$

where $A_U = U A_\mu U^{-1} + \frac{i}{g} U \partial_\mu U^{-1}$, and $\mathcal{D}U$ is the functional generalization of the group invariant measure dU for SU(N) which has the property that $\int dU = \int d(U'U)$ for any fixed element U' of SU(N). We will not have to concern ourselves with the detailed form of this measure since it just goes into the overall infinite factor which we absorb into the normalization of the path integral. But we do have to compute the determinant, which looks formidable. Fortunately though, we can compute it using the infinitesimal form of the gauge transformation:

$$A_U = (1 + ig\omega^a T^a) A_\mu (1 - ig\omega^a T^a) + \frac{i}{g}(-ig)\partial_\mu \omega^a T^a$$

$$\longrightarrow \delta A = ig\omega^a A_\mu^b [T^a, T^b] + \partial_\mu \omega^a T^a$$

$$= -g f^{abc} \omega^a A_\mu^b T^c + \partial_\mu \omega^a T^a \,. \tag{13.26}$$

Therefore

$$\delta A_\mu^c = \partial_\mu \omega^c - g f^{abc} \omega^a A_\mu^b \,. \tag{13.27}$$

Let's consider the form of the gauge condition for covariant gauges. In analogy to QED, we would like to take

$$f(A) = \partial^\mu A_\mu = \partial^\mu A_\mu^a T_a = 0 \,. \tag{13.28}$$

We see that there are really separate conditions for each gauge boson, so our gauge condition should have a color index as well: $f \to f^a$. Then the FP determinant takes the form

$$\Delta_{\text{FP}}[A] = \det \frac{\partial f^a}{\partial A_\mu^c} \frac{\delta(\delta A_\mu^c)_x}{\delta(\omega_y^d)} = \det \left| \partial_\mu (\delta_{ad} \partial_\mu - g f^{dba} A_\mu^b) \delta(x - y) \right| \,. \tag{13.29}$$

Unlike in QED, the nonAbelian nature of the theory has introduced field-dependence into the FP determinant, so we will no longer be able to ignore it.

The problem now is how to account for this new factor in a way which we know how to compute. We know from previous experience though that an anticommuting field produces a functional determinant when we perform its path integral. Therefore it must be possible to rewrite the FP determinant as a path integral over some new fictitious fields c_a, $\bar{c}_a$ which we call *ghosts*,

$$\Delta_{\text{FP}}[A] = \int \mathcal{D}\bar{c}\,\mathcal{D}c\, e^{i S_{\text{ghost}}} , \tag{13.30}$$

where

$$S_{\text{ghost}} = \int d^4x\, \bar{c}_a \delta_{ad} \left(-\partial^\mu \partial_\mu + g f^{dba} \partial^\mu A_\mu^b\right) c_d . \tag{13.31}$$

Similarly to our derivation of the self interactions of the gauge bosons, we can obtain the Feynman rules for the ghost propagator and the ghost-ghost-gluon vertex,

$$a \text{----} \blacktriangleleft \text{----} b = i \frac{\delta_{ab}}{p^2};$$

$$= -g f^{abc} p$$

$$\tag{13.32}$$

The ghosts are fermionic fields, whose path integral produces the determinant with a positive power, unlike bosons which give a negative power. Loops of the ghost fields will therefore come with a minus sign, just like those of fermions. On the other hand, their propagator looks like that of a scalar field. Therefore the ghosts violate the spin-statistics theorem, which says that a scalar particle should be quantized as a boson. However, the ghosts are not physical fields; their purpose is to cancel out the unphysical contributions of the longitudinal and timelike polarizations of the gauge bosons. In QED the unphysical polarizations never contributed to physical amplitudes in the first place, so it was not necessary to cancel them. But in Yang-Mills theory, the unphysical states participate in the interaction vertices, so their effects must be explicitly counteracted. This will always come about through loops of the ghost fields; at tree level they are not necessary. To appreciate this important fact, let's look at the issue of unitarity of the S-matrix in greater detail.

13.3 Unitarity of the S-Matrix

Conservation of probability in quantum mechanical scattering theory is expressed by the unitarity of the scattering matrix,

$$S^\dagger S = 1 \,. \tag{13.33}$$

Let us change notation slightly to that of Ref. [14] to express the relationship between the S matrix, the T matrix, and the transition amplitude $\mathcal{M}$:

$$S = 1 + iT \,, \tag{13.34}$$

where 1 is the identity operator. Then unitarity says that

$$S^\dagger S = 1 + i(T - T^\dagger) + T^\dagger T = 1 \,. \tag{13.35}$$

The matrix elements of T for a $2 \to n$ scattering process, for example, are related to $\mathcal{M}$ by

$$\langle \mathbf{p}_1 \cdots \mathbf{p}_n | T | \mathbf{q}_1 \mathbf{q}_2 \rangle = (2\pi)^4 \delta^{(4)}(q_1 + q_2 - p_1 \cdots - p_n) \mathcal{M}(q_i \to p_i);$$
$$\langle \mathbf{q}_1' \mathbf{q}_2' | T^\dagger | \mathbf{p}_1 \cdots \mathbf{p}_n \rangle = (2\pi)^4 \delta^{(4)}(q_1' + q_2' - p_1 \cdots - p_n) \mathcal{M}^*(q_i' \to p_i) \,. \tag{13.36}$$

One of the most common uses of unitarity is to take the matrix element of (13.35) between a pair of two-particle states (though we could equally well consider any number of particles):

$$-i\left(\langle \mathbf{q}_1' \mathbf{q}_2' | T | \mathbf{q}_1 \mathbf{q}_2 \rangle - \langle \mathbf{q}_1' \mathbf{q}_2' | T^\dagger | \mathbf{q}_1 \mathbf{q}_2 \rangle \right) = \langle \mathbf{q}_1' \mathbf{q}_2' | T^\dagger T | \mathbf{q}_1 \mathbf{q}_2 \rangle$$
$$= \sum_i \langle \mathbf{q}_1' \mathbf{q}_2' | T^\dagger | i \rangle \langle i | T | \mathbf{q}_1 \mathbf{q}_2 \rangle \,. \tag{13.37}$$

The explicit form for the complete set of states inserted in (13.37) is

$$\sum_i |i\rangle\langle i| = \sum_n \left(\prod_{i=1}^n \int \frac{d^3 p_i}{(2\pi)^3} \frac{1}{2E_i} \right) |\mathbf{p}_1 \cdots \mathbf{p}_n\rangle\langle \mathbf{p}_1 \cdots \mathbf{p}_n|$$
$$= \sum_n \left(\prod_{i=1}^n \int \frac{d^4 p_i}{(2\pi)^3} \delta(p_i^2 - m_i^2)\theta(p_i^0) \right) |\mathbf{p}_1 \cdots \mathbf{p}_n\rangle\langle \mathbf{p}_1 \cdots \mathbf{p}_n|$$
$$\equiv \sum_f \int d\Pi_f |f\rangle\langle f| \,. \tag{13.38}$$

When we eliminate T in favor of $\mathcal{M}$, this equation takes the form

$$-i\left(\mathcal{M}(q_i \to q_i') - \mathcal{M}^*(q_i' \to q_i) \right) = \sum_f \int d\Pi_f \mathcal{M}(q_i \to p_i) \mathcal{M}^*(q_i' \to p_i) \,. \tag{13.39}$$

Equation (13.39) clearly requires the existence of an imaginary part of the amplitude for the l.h.s. to be nonzero. For example, at tree level in $\lambda\phi^4$ theory, we have

Fig. 13.2 Analytic structure
of the $2 \rightarrow 2$ amplitude in
$\lambda \phi^4$ theory in the complex
q^2 plane

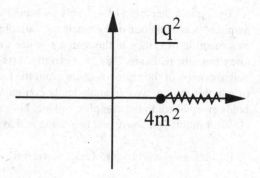

$\mathcal{M} = \lambda$, so the l.h.s. vanishes. At one loop however, an imaginary part can occur at certain values of the external momenta. Recall the result (6.2):

$$\mathcal{M}(2 \rightarrow 2) = \sum_{q^2 = s,t,u} \frac{\lambda^2}{32\pi^2} \left[\ln(\Lambda^2/m^2) + 1 - \sqrt{1 - 4m^2/q^2} \ln \left(\frac{\sqrt{1 - 4m^2/q^2} + 1}{\sqrt{1 - 4m^2/q^2} - 1} \right) \right].$$
(13.40)

This expression is rather ambiguous because it does not tell us which branch we should take when the argument of the square root becomes negative. This can be determined by more carefully comparing to the original expressions before carrying out the Feynman parameter integral, and paying attention to the $i\epsilon$ factors. Nevertheless, it can be seen that the function $z \ln((z + 1)/(z - 1))$ is purely real when z is purely imaginary, which corresponds to $q^2 < 4m^2$, whereas it has both real and imaginary parts when z is real, i.e., when $q^2 > 4m^2$. This function has the analytic behavior shown in Fig. 13.2: a branch cut develops in the complex q^2 plane at $4m^2$.

The significance of the value $4m^2$ can be seen by considering the special case where $q_i' = q_i$ in (13.37). Then $\mathcal{M}(q_i \rightarrow q_i)$ is the *forward scattering amplitude*, and the r.h.s. of (13.37) is proportional to the total cross section for $q_1, q_2 \rightarrow$ anything, since the integrals are of the same form as the phase space for the final states. This is the **optical theorem**,

$$2\text{Im}\mathcal{M}(q_i \rightarrow q_i) = \sqrt{(q_1 \cdot q_2)^2 - m_1^2 m_2^2} \; \sigma_{\text{tot}}(q_1, q_2 \rightarrow \text{anything}).$$
(13.41)

The factor $\sqrt{(q_1 \cdot q_2)^2 - m_1^2 m_2^2}$ is often written as $2E_{\text{cm}} p_{\text{cm}} = \sqrt{s} \, p_{\text{cm}}$, and is related to the relative velocity of the colliding particles by $v_{\text{rel}} = \sqrt{(q_1 \cdot q_2)^2 - m_1^2 m_2^2} / (E_1 E_2)$. Notice that we are allowed to evaluate the l.h.s. of the equation off-shell, for example at vanishing external momenta. But in the case of $\lambda \phi^4$ theory, the cross section does not exist for these unphysical values of the initial momenta, so the r.h.s. is zero. Only when we have enough energy in the initial state to produce at least two real particles does the scattering cross section begin to be nonzero. This requires a minimum value of $s = (m + m)^2 = 4m^2$.

The optical theorem (13.41) tells us something about the analytic properties of amplitudes as a function of the kinematic variables. For example if we consider $\mathcal{M}(s)$ as a function of s, then in the region $s < 4m^2$, we can analytically continue $\mathcal{M}(s)$ away from the real axis. Using the identity $\mathcal{M}(s) = (\mathcal{M}(s^*))^*$ which holds in some neighborhood of the region $s < 4m^2$, and the fact that $\mathcal{M}(s)$ is an analytic function of s in this region, we can analytically continue to deduce that $\mathcal{M}(s) = (\mathcal{M}(s^*))^*$ holds everywhere in the complex s plane. Now if we go to the region $s > 4m^2$ and consider points just above and below the real axis, $s \pm i\epsilon$, we see that

$$\mathrm{Re}\mathcal{M}(s + i\epsilon) = \mathrm{Re}\mathcal{M}(s - i\epsilon); \quad \mathrm{Im}\mathcal{M}(s + i\epsilon) = -\mathrm{Im}\mathcal{M}(s - i\epsilon). \quad (13.42)$$

There is a *discontinuity* in the imaginary part across the cut.

The $i\epsilon$ prescription of the Feynman propagators give us a nice way of computing the imaginary parts of diagrams. We use the identity

$$\frac{1}{x + i\epsilon} = \frac{x - i\epsilon}{x^2 + \epsilon^2} = P\left(\frac{1}{x}\right) - i2\pi\delta(x), \quad (13.43)$$

where $P(1/x)$ denotes the principal value of $1/x$. In the case of a bosonic propagator, this gives

$$\frac{1}{p^2 - m^2 + i\epsilon} = P\left(\frac{1}{p^2 - m^2}\right) - i2\pi\delta(p^2 - m^2). \quad (13.44)$$

One can show that the imaginary part of the amplitudes we have considered is due to the second term in (13.44). One should not be misled into thinking that the imaginary part of the product of all propagators as per (13.44) gives the imaginary part of the amplitude however. The optical theorem tells us that *all* the internal lines must be on shell. Moreover, the delta function in (13.44) gets contributions from both positive and negative energy states, whereas the optical theorem tells us that $\mathrm{Im}\mathcal{M}$ is associated with positive energy particles only. Therefore it is not trivial to show the link between (13.44) and $\mathrm{Im}\mathcal{M}$. The result, which I shall not prove, but which follows from the optical theorem, is known as the Cutkosky rules: to find $\mathrm{Im}\mathcal{M}$, make the following replacement on all the internal lines which are going on shell:

$$\frac{i}{p^2 - m^2 + i\epsilon} \rightarrow 2\pi\theta(p^0)\delta(p^2 - m^2). \quad (13.45)$$

Now let's relate this ideas to gauge theories. Our concern is that unphysical states of the gauge boson must not contribute to the imaginary part of an amplitude. If they did, then unitarity would require them to also appear as external states in tree-level processes. For example, the optical theorem requires the relation depicted in Fig. 13.3 for scattering of two gauge bosons. Notice that this process did not arise in the Abelian theory. Explicit calculations show that in fact this relationship fails because the longitudinal and timelike polarizations of the gauge fields do contribute to the imaginary part of the loop amplitude. This is why the Faddeev-Popov ghosts

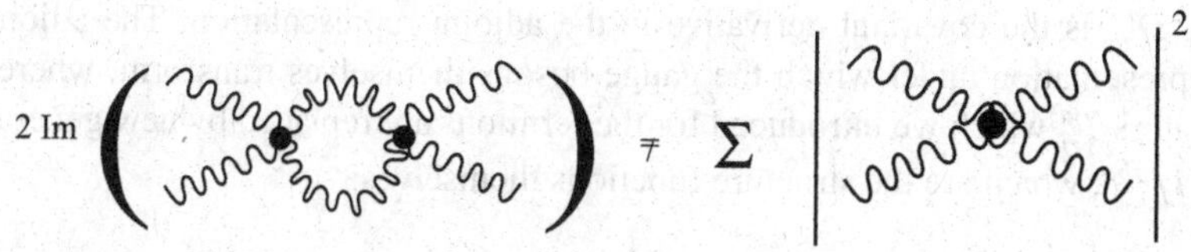

Fig. 13.3 Apparent failure of unitarity in Yang-Mills theory

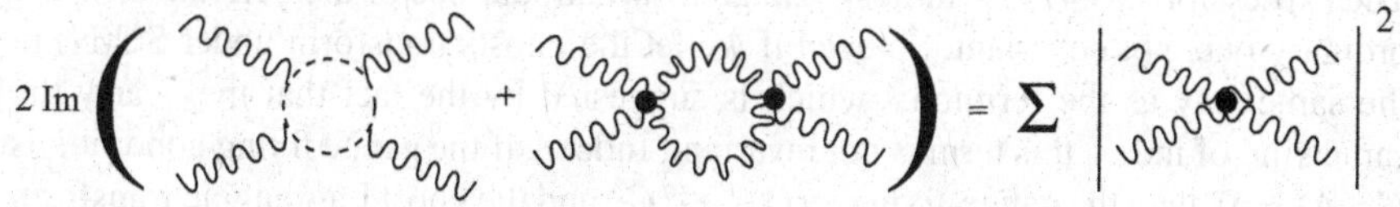

Fig. 13.4 Restoration of unitarity by Faddeev-Popov ghost loop contribution

are necessary in the nonAbelian theory: their contribution subtracts the unphysical one from the loop amplitudes (Fig. 13.4).

13.4 Loop Structure of Non Abelian Gauge Theory

Now that have shown the importance of the ghosts for unitarity, let's return to the renormalization of the nonAbelian theory. Now there are more counterterms, because of the new cubic, quartic, and ghost interactions of the gauge bosons. These new terms also make the analog of the QED Ward identities, called the Slavnov-Taylor identities, much more complicated than in QED. First, let's consider the form of the renormalized Lagrangian. If we write out explicitly the new gauge interactions, it has the form

$$\mathcal{L}_{\text{ren}} = Z_2 \bar{\psi} i \partial\!\!\!/ \psi - m Z_m \bar{\psi}\psi + g\mu^\epsilon Z_1 \bar{\psi} A\!\!\!/ \psi + \frac{1}{4} Z_3 (\partial_\mu A_\nu - \partial_\nu A_\mu)(\partial^\mu A^\nu - \partial^\nu A^\mu)$$

$$- Z_4 g\mu^\epsilon f^{abc} A_a^\mu A_b^\nu \partial_\mu A_\nu^c + \frac{1}{2} Z_5 g^2 \mu^{2\epsilon} f^{abc} f^{ade} A_\mu^b A_\nu^c A_d^\mu A_e^\nu + \frac{1}{2} Z_\alpha (\partial_\mu A^\mu)^2$$

$$+ Z_6 \partial_\mu \bar{c}^a \partial^\mu - \frac{1}{2} Z_7 g\mu^\epsilon f^{abc} A_\mu^c \bar{c}^a \overleftrightarrow{\partial}_\mu c^b - \frac{1}{2} Z_8 g\mu^\epsilon f^{abc} \partial^\mu A_\mu^c \bar{c}^a c^b . \tag{13.46}$$

Note that we have suppressed color indices in some places. In the case of QED, we were able to simplify some calculations due to the relation $Z_1 = Z_2$ which came from the Ward identities. In QCD this is no longer possible; the ghost Lagrangian presents a new source of gauge noninvariance when we try to carry out the same path integral argument that led to the U(1) Ward identities. The problem can be seen by looking at the form of the ghost Lagrangian; at tree level it can be written as

$$\mathcal{L}_{\text{ghost}} = \bar{c}^a \partial_\mu D_{ab}^\mu c^b , \tag{13.47}$$

where D^{μ}_{ab} is the covariant derivative in the adjoint representation. The adjoint is the representation under which the gauge bosons themselves transform, where the generators T^a_{ij} which we introduced for the fermions are replaced by new generators $t^a_{bc} = if^{abc}$, which are the structure functions themselves:

$$D^{\mu}_{ab}c^b = \left(\delta_{ac}\partial_{\mu} + gf^{abc}A^b_{\mu}\right)c^b .\tag{13.48}$$

The expression (13.47) is "almost" gauge invariant, but one of the derivatives is the ordinary one, not covariant. So even if we let the ghosts transform under SU(N) in the same way as the fermions, which is suggested by the fact that they carry the same kind of index, this term is not invariant. Indeed, if the gauge fixing condition is $G^a(A) = 0$, then the gauge-fixing term $\frac{1}{2\alpha}G^aG^a$ and the ghost Lagrangian transform to themselves plus the following piece that expresses the lack of invariance under an infinitesimal gauge transformation ω^a:

$$\delta\mathcal{L}_{\text{g.f.+ghost}} = \frac{1}{\alpha}\frac{\delta G^a}{\delta A^b_{\mu}}D^{bc}_{\mu}\omega^c + \delta\bar{c}^a\frac{\delta G^a}{\delta A^b_{\mu}}(D_{\mu}c)^b + \bar{c}^a\frac{\delta G^a}{\delta A^b_{\mu}}\delta(D_{\mu}c)^b .\tag{13.49}$$

Notice that we are not forced to assume that the ghosts transform in exactly the same way as the fermions in (13.49), so we have left $\delta\bar{c}$ and δc unspecified. In fact, it is possible to choose them in a clever way that makes $\delta\mathcal{L}_{\text{g.f.+ghost}}$ vanish. This was discovered by Becchi, Rouet and Stora and around the same time by Tyutin, and is known as the BRST symmetry. This elegant technique is not only helpful for proving the renormalizability of Yang-Mills theory, but also has applications in string theory. The symmetry requires the gauge transformation to depend on the ghost itself:

$$\omega^a = \zeta c^a; \qquad \delta\bar{c}^a = -\frac{1}{\alpha}G^a\zeta ,\tag{13.50}$$

where $\zeta(x)$ must also be a Grassmann-valued field so that ω^a can be a c-number (an ordinary commuting number). The motivation for this choice is that the first two terms of (13.49) cancel each other. Now we only need to define δc^a in such a way that $\delta(D_{\mu}c)^b$ vanishes. One might think that we no longer have the freedom to do this since the transformation of $\bar{c}^a$ has been defined, but in fact it is kosher to treat $\bar{c}$ and c as independent fields in the path integral. (This can be proven by explicitly writing everything in terms of the real and imaginary parts of c and $\bar{c}$.) It can be verified that the magic form of the transformation of c is

$$\delta c^b = \frac{1}{2}\zeta f^{bce}c^c c^e .\tag{13.51}$$

We now have a new continuous symmetry of the gauge-fixed Lagrangian which can be used to find relations between amplitudes in the nonAbelian theory. In terms of the effective action, these relations (the Slavnov-Taylor identities) can be written by introducing new source terms which couple to the changes that are induced in the fields under the BRST transformations:

$$\mathcal{L}_{\text{source}} = \tau_\mu^a (D_\mu c)^a/g + u^a f^{abc} c^b c^c/2 + \bar{\lambda} c^a T^a \psi + \bar{\psi} T^a c^a \lambda \qquad (13.52)$$

The new sources τ_μ^a, u^a, $\bar{\lambda}$ and λ are analogous to the source terms $J^\mu A_\mu$ for the gauge field or $J\phi$ for a scalar field. Notice that λ and $\bar{\lambda}$ are fermionic source terms since the couple to the change in the fermionic field ψ. The utility of these source terms is that if we compute the effective action as a function of the new sources and functionally differentiate it with respect to the sources, we get the expressions for the changes in the classical fields with respect to the BRST transformations. The Slavnov-Taylor identities can then be written as

$$\int d^4x \left(\frac{\delta\Gamma}{\delta A_\mu^b} \frac{\delta\Gamma}{\delta \tau_b^\mu} - \frac{1}{\alpha g} \partial_\mu A_b^\mu \frac{\delta\Gamma}{\delta \bar{c}^b} - \frac{\delta\Gamma}{\delta c^b} \frac{\delta\Gamma}{\delta u^b} + i\frac{\delta\Gamma}{\delta\psi} \frac{\delta\Gamma}{\delta\bar{\lambda}} - i\frac{\delta\Gamma}{\delta\bar{\psi}} \frac{\delta\Gamma}{\delta\lambda} \right) = 0.$$
$$(13.53)$$

Although it may not be obvious, these identities actually reduce to the Ward identities in the U(1) case. Essentially, each functional relation contained in the Ward identities is also contained in the Slavnov-Taylor relations, except multiplied by a power of the ghost fields. So instead of looking for a term which is of order ψ^0, $\bar{\psi}^0$, A^1, we would look for the same term but also of order c^1 in the ghost field. For example, $\frac{\delta\Gamma}{\delta \tau_b^\mu}$ gives us $\partial_\mu c^b$ at lowest order in perturbation theory, whereas $\frac{\delta\Gamma}{\delta \bar{c}^b}$ gives us $\partial^2 c^b$. Thus the terms of order A^1, c^1 are

$$\int d^4x \left(\frac{\delta\Gamma}{\delta A_\mu^b} \partial_\mu c^b - \frac{1}{\alpha g} \partial_\mu A_b^\mu \partial^2 c^b \right), \qquad (13.54)$$

which agrees with the QED Ward identity (12.42).

These identities are still quite complicated (but they would have been even worse had we tried to work with the broken gauge symmetry rather than the BRST transformations). They can be used to prove the following relations between the renormalization constants:

$$\frac{Z_1}{Z_2} = \frac{Z_4}{Z_3} = \frac{\sqrt{Z_5}}{\sqrt{Z_3}} = \frac{Z_7}{Z_6} = \frac{Z_8}{Z_6}. \qquad (13.55)$$

In particular, we no longer have $Z_1 = Z_2$, due to the new complications from the ghosts. This means that we can no longer simplify the relation between the bare and renormalized couplings,

$$g_0 = g\mu^\epsilon \frac{Z_1}{Z_2 Z_3^{1/2}} \qquad (13.56)$$

to get away with computing only Z_3 as we did in QED. Rather, we are forced to compute all the diagrams contributing to the fermion vertex and wave function renormalization, in addition to the new ghost loop contributions to the vacuum polarization. There is one exception to this statement: in the *axial gauge*

$$n_\mu A_b^\mu = 0 \tag{13.57}$$

for some fixed vector n_μ, the ghosts have no interaction with the gauge bosons, and they can be ignored once again. This can be seen by considering the form of the determinant:

$$\Delta_{\text{FP}}[A] = \det \frac{\partial f^a}{\partial A_\mu^c} \frac{\delta(\delta A_\mu^c)_x}{\delta(\omega_y^d)} = \det \left| n_\mu(\delta_{ad}\partial_\mu - g f^{dba} A_\mu^b)\delta(x-y)\right|$$

$$= \det \left| n_\mu(\delta_{ad}\partial_\mu)\delta(x-y)\right| . \tag{13.58}$$

We used the fact that this determinant is multiplied by $\delta[n_\mu A^\mu]$ in the path integral to set $n_\mu A_a^\mu = 0$. The problem with this gauge is that the gauge boson propagator is so complicated that it hardly makes life easier,

$$D_{\mu\nu}^{ab}(p) = -\frac{\delta_{ab}}{p^2}\left(\eta_{\mu\nu} - \frac{1}{n\cdot p}(n_\mu p_\nu + n_\nu p_\mu) - \frac{p_\mu p_\nu}{(n\cdot p)^2}(\alpha p^2 - n^2)\right) . \tag{13.59}$$

There is a way we could have guessed the relations (13.55) without going through all the machinery of the Slavnov-Taylor relations. When we rewrite the renormalized Lagrangian in terms of the bare fields, it takes the form

$$\mathcal{L}_{\text{ren}} = \bar\psi i\partial\!\!\!/\psi - m\bar\psi\psi + g_0\bar\psi A\!\!\!/\psi + \frac{1}{4}(\partial_\mu A_\nu - \partial_\nu A_\mu)(\partial^\mu A^\nu - \partial^\nu A^\mu)$$

$$- g_0^{(1)} f^{abc} A_a^\mu A_b^\nu \partial_\mu A_\nu^c + \frac{1}{2}(g_0^{(2)})^2 f^{abc} f^{ade} A_\mu^b A_\nu^c A_d^\mu A_e^\nu + \frac{1}{2\alpha_0}(\partial_\mu A^\mu)^2$$

$$+ \partial_\mu \bar c^a \partial^\mu - \frac{1}{2}g^{(3)} f^{abc} A_\mu^c \bar c^a \overleftrightarrow{\partial}_\mu c^b - \frac{1}{2}g^{(4)} f^{abc} \partial^\mu A_\mu^c \bar c^a c^b . \tag{13.60}$$

For clarity I have omitted the subscript "0" on all the bare fields in (13.60). The various bare couplings are related to the renormalized one by

$$g_0 = g\mu^\epsilon \frac{Z_1}{Z_2 Z_3^{1/2}} \tag{13.61}$$

$$g_0^{(1)} = g\mu^\epsilon \frac{Z_4}{Z_3^{3/2}} \tag{13.62}$$

$$g_0^{(2)} = g\mu^\epsilon \frac{Z_5^{1/2}}{Z_3} \tag{13.63}$$

$$g_0^{(3)} = g\mu^\epsilon \frac{Z_7}{Z_6 Z_3^{1/2}} \tag{13.64}$$

$$g_0^{(4)} = g\mu^\epsilon \frac{Z_8}{Z_6 Z_3^{1/2}} \tag{13.65}$$

If all these bare couplings are equal to each other, then we can rewrite the whole Lagrangian in terms of covariant derivatives and field strengths, so that its form is gauge invariant except for the gauge fixing and ghost term. And even though the ghost term is not invariant, it takes its original tree level form, $\bar{c}^a \partial_\mu (D^\mu c)_a$. These equalities are insured if the renormalization constants obey the relations (13.55).

13.5   Beta Function of Yang-Mills Theory

If we want to show that QCD is asymptotically free, the above arguments indicate that we have to compute the divergent parts of all of the following diagrams. For Z_2, the fermion wave function renormalization, there are no new kinds of diagrams; the self-energy diagram is the same as Fig. 12.2 for QED. The only difference is that there is a group theory factor from the vertices:

$$= T_{ij}^a T_{jk}^a \times (\text{QED result})$$

$$= \frac{N^2 - 1}{2N} \delta_{ij} \times (\text{QED result}),$$

(13.66)

where we used the result (13.11) for $T_{ij}^a T_{jk}^a$.

For the vertex correction, we have the diagram similar to Fig. 12.4 for QED, but in addition the new contribution of Fig. 13.5. I will not go through the computation of this new diagram. However, it is rather straightforward to compute the one which is similar to QED,

Fig. 13.5 New vertex correction in nonAbelian theory

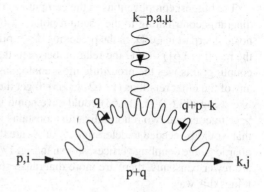

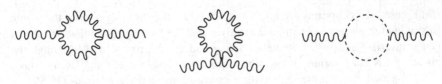

Fig. 13.6 New contributions to the vacuum polarization in nonAbelian theory

$$
\vcenter{\hbox{\includegraphics{fig}}} = T_{il}^b T_{lm}^a T_{mj}^b \times (\text{QED result})
$$

(13.67)

We can write the group theory factor as

$$
T^b T^a T^b = [T^b, T^a] T^b + T^a T^b T^b = i f^{bac} T^c T^b + \frac{N^2 - 1}{2N} T^a
$$

$$
= \frac{i}{2} f^{bac} [T^c, T^b] + \frac{N^2 - 1}{2N} T^a
$$

$$
= -\frac{1}{2} f^{bac} f^{cbd} T^d + \frac{N^2 - 1}{2N} T^a
$$

$$
= -\frac{1}{2} N T^a + \frac{N^2 - 1}{2N} T^a .
$$

(13.68)

We used the identity (13.12) to obtain the last line. (The minus sign in $-\frac{1}{2} N T^a$ is erroneously missing in Ramond (6.52) and (6.53).)

The biggest complications in the computation of the beta function are all the new diagrams contributing to the vacuum polarization, shown in Fig. 13.6. These have no counterpart in QED. In the preceding description, we assumed that we would use the result (13.61) to find the relation between the bare and the renormalized gauge coupling; this was the procedure most analogous to QED. But we could have used any of the other relations (13.62–13.65) to get the same result. For example, it looks like using (13.62) or (13.63) could save some work because then we would only have to compute two renormalization constants instead of three. The new diagrams that would be needed to determine Z_4 or Z_5 are the renormalization of the cubic and quartic gauge coupling vertices, shown in Fig. 13.7. It is clear that, even though there are fewer constants, there are more diagrams, so we don't save any work by doing things this way.

It is remarkable that these very different and complicated ways of computing the renormalized coupling give the same answer, thanks to the gauge symmetry. The

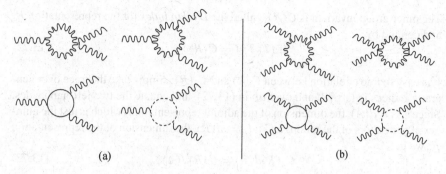

Fig. 13.7 Contributions to renormalization of **a** cubic and **b** quartic gauge couplings

result for the beta function can be expressed in a way which is applies to gauge groups which are more general than SU(N),

$$\beta(g) = -\frac{g^3}{16\pi^2}\left(\frac{11}{3}C_2(A) - \frac{4}{3}n_f C(f) - \frac{1}{3}C(s)\right), \tag{13.69}$$

where n_f is the number of flavors of fermions, and the numbers C_2 and C are group theoretic factors which are a function of the representation of the gauge group under which the argument transforms. I have also shown the additional effect we would get from enlarging our theory to have complex scalar fields which are charged under the gauge group. They contribute just like fermions, except we have assumed the fermions are Dirac particles with 4 degrees of freedom, and each fermion carries as much weight as 4 complex scalars. $C_2(A)$ is the *quadratic Casimir invariant* for the gauge fields, which transform in the adjoint representation. This invariant is defined by the identity

$$\sum_a T_R^a T_R^a \equiv C_2(R)I_R, \tag{13.70}$$

where T_R^a is the group generator in the representation R and I_R is the unit matrix which has the same dimensions as the generators. Even if you have not taken a course in group theory, you are familiar with the fact that the matrix elements of the angular momentum operators form matrices whose dimension is $(2l + 1) \times (2l + 1)$ for states with angular momentum l. It so happens that for the rotation group SO(3), since there are three generators, the adjoint representation coincides with the fundamental representation, since 3-vectors are the fundamental objects which are transformed by SO(3). But in general the adjoint and fundamental representations are different. From the identities (13.11) and (13.12) we deduce that for SU(N),

$$C_2(R) = \frac{N^2 - 1}{2N}, \qquad C_2(A) = N. \tag{13.71}$$

The other group invariant is $C(R)$, called the *Dynkin index* for the representation R, and defined by

$$\text{tr}\left(T_R^a T_R^b\right) = C(R)\delta_{ab}. \tag{13.72}$$

It is easy to find a relation between $C(R)$ and $C_2(R)$; simply take the trace over generator indices in (13.70) and over ab in (13.72), and equate the two left-hand sides. Since $\delta_{ab} = d(A)$, the dimension of the adjoint representation (which is just the number of generators of the group) and $\text{tr}\, I_R = d(R)$, the dimension of the representation R, we get

$$C_2(R)d(R) = C(R)d(A). \tag{13.73}$$

For fermions in the fundamental representation of SU(N) this gives

$$C(f) = \frac{N^2 - 1}{2N} \times N/(N^2 - 1) = \frac{1}{2}. \tag{13.74}$$

Putting these together and ignoring the effect of possible scalar fields, we get the beta function in SU(N),

$$\beta(g) = -\frac{g^3}{16\pi^2}\left(\frac{11}{3}C_2(A) - \frac{4}{3}n_f C(f)\right) = -\frac{g^3}{16\pi^2}\left(11 - \frac{2}{3}n_f\right). \tag{13.75}$$

It is clear that the new term which makes this negative (if n_f is not too large!) is due to the gauge bosons. Roughly speaking, the factor $C_2(A)$ counts how many gauge bosons are going around the new loops due to gauge boson self-interactions, whereas $n_f C(f)$ counts the contributions from fermion loops. If we set $C_2(A) = 0$ we can recover our old QED result by taking $n_f = 1$ and identifying $g = \sqrt{2}e$. The reason for the factor of 2 is the conventional way in which generators of SU(N) are normalized: $\text{tr}\, T_a T_b = \frac{1}{2}\delta_{ab}$. In QED we take $T = 1$, not $T = 1/\sqrt{2}$.

Although the fermions weaken asymptotic freedom, in the real world of QCD there are only 6 flavors of quarks. We would need 17 flavors to change the sign of the beta function.

13.6   Heuristic Explanation of Asymptotic Freedom

We see from (13.75) that there is a competition between the gauge bosons and the fermions for determining the sign of the beta function. If we add enough flavors of fermions, they win and spoil the asymptotic freedom of the theory. We know that fermions have this effect in QED, because their vacuum fluctuations screen the bare charge of the electron at long distances, causing the coupling to become weaker at low energy. Apparently the gauge bosons in the nonAbelian theory have the effect of *antiscreening* a bare color charge. Can we understand why this happens? Peskin has a nice explanation (Sect. 16.7 of [14]), but I will instead give one from [7], originally found in papers of Nielsen (1980) and Hughes (1981).

In the following discussion we will make use of our intuition from ordinary electromagnetism. In the Yang-Mills theory it is also possible to talk about electric and magnetic fields,

$$E_i^a = F_{0i}^a, \quad B_i^a = \frac{1}{2}\epsilon_{ijk}F_{jk}^a, \tag{13.76}$$

but now there are $N^2 - 1$ components labeled by the group index. Without the new self-interactions of the gauge bosons, this would just be $N^2 - 1$ copies of QED. So for the interaction of the gauge bosons with fermions, at lowest order in the coupling, we can use our intuition from QED. Of course all the new effects will come from the gauge bosons' interactions with themselves.

We can think of the beta function as telling us about the energy dependence of the dielectric constant for a bare charge, due to fluctuations of the vacuum:

$$g^2(E) = g_0^2/\epsilon(E)$$

$$= \frac{g_0^2}{1 - 2b_0 g_0^2 \ln(E/\Lambda)}, \tag{13.77}$$

where I am using E for the renormalization scale and Λ for the cutoff. The constant b_0 is related to the beta function by

$$\beta(g) = b_0 g^3 + \cdots \tag{13.78}$$

We can thus identify the "running electric permeability" as

$$\epsilon(E) = 1 - 2b_0 g_0^2 \ln(E/\Lambda). \tag{13.79}$$

In QED we know that $\epsilon(E) > 1$ always, corresponding to screening. Apparently in QCD we have antiscreening, corresponding to $\epsilon(E) < 1$.

Instead of thinking about the electric polarizability of the vacuum, we could also consider the magnetic permeability μ, since

$$\mu\epsilon = c^2 = 1. \tag{13.80}$$

Therefore, if we can understand why $\mu > 1$ for nonAbelian theories, this will explain why $\epsilon < 1$. In fact, the coefficient of the beta function is very directly related to the *magnetic susceptibility*,

$$\chi(E) = 1 - 1/\mu(E) = -2b_0 g_0^2 \ln(\Lambda/E). \tag{13.81}$$

Thus the sign of the beta function is the negative of the sign of the susceptibility.

We are quite familiar with an effect that leads to $\mu > 1$ in ordinary electromagnetism: place a piece of iron inside a solenoid, and it increases the magnetic field by a factor $\mu > 1$, due to the fact that the spin magnetic moments of the iron atoms like to align with the external field, thus increasing the total field. This effect is known as *paramagnetism*. However, there is another effect which leads to $\mu < 1$: if a charged particle undergoes cyclotron motion around the magnetic field lines, it creates a current loop whose magnetic dipole moment *opposes* the direction of the external field and causes a decrease in the net field, leading to $\mu < 1$. This effect is called *diamagnetism*. Clearly, if we have quantum fluctuations of virtual fermions and gauge bosons in the vacuum (in an external magnetic field), both effects are possible. It becomes a quantitative question of which one is stronger. Apparently, diamagnetism must be the stronger effect for fermions, since otherwise we would get $\mu > 1$ and $\epsilon < 1$ even in QED. So if this is the correct explanation, it must be because paramagnetism is the bigger effect for the nonAbelian gauge bosons. In QED they were irrelevant because of the fact that photons don't couple to themselves, so they can't act as a source of magnetic field. But in QCD, a vacuum fluctuation of the gauge bosons can act as a source of magnetic field, since the gauge field equation of motion (ignoring fermionic sources) is

$$\partial^\mu F_{\mu\nu} + ig[A^\mu, F_{\mu\nu}] = \partial^0 F_{0\nu} + \partial^i F_{\mu i} + ig[A^0, F_{0\nu}] + ig[A^i, F_{i\nu}] = 0; \tag{13.82}$$

this is the generalization of Maxwell's equations to the nonAbelian case. To find the magnetic part, take $\nu = j$,

$$\partial_0 E_j + \partial^i \epsilon_{jik} B_k + ig[A^0, E_j] + ig[A^i, \epsilon_{jik} B_k] = 0. \tag{13.83}$$

We see that if the background gauge fields have $E_j = 0$ but $B_k \neq 0$, then a vacuum fluctuation of A can give rise to a source of color magnetic field due to the last term.

To compute the beta function from this argument, we will use another result from classical electromagnetism: the self-energy of a magnetic field configuration in a magnetic medium is given by

$$U = -\frac{1}{2}\chi B^2. \tag{13.84}$$

We will compute the correction to U due to the fluctuations of the vacuum, and from this infer the beta function. The argument is similar in spirit to our intuitive explanation of the axial anomaly. We want to see how the energy levels of the quantum states of the theory are affected by a background magnetic field, due to the interactions of the states with the field.

First let's compute the diamagnetic contribution. If the B field is in the $\hat{z}$ direction, then the energy eigenstates of a charged particle are the Landau levels,

$$E^2 = p_z^2 + m^2 + (2n+1)eB, \tag{13.85}$$

with $n \geq 0$, and $e = g/\sqrt{2}$, according to the observation below (13.75). Here we should imagine that the B field is pointing in one particular direction in the adjoint representation color space, say a, so the fermion or gauge boson which has the Landau levels is some linear combination associated with the generator T^a. Because of gauge invariance, it should not matter which generator we choose, so for convenience consider one of the diagonal generators, like T_3 for the fermions. Then we see that fermions with color indices ψ_i, $i = 1, 2$ couple to this external field, while $i = 3$ does not. Since the effect we are looking for in (13.84) is quadratic in B, it will not matter that $i = 1$ and $i = 2$ couple with different signs. According to this argument, the effect we are looking for will be proportional to the trace of the square of the generator, and this is the same for all generators, so this supports our claim that the choice of the B field's direction in color space will not affect our answer. Note that the extra factor of two we get here will cancel the factors of $1/\sqrt{2}$ which arise from the relation $e = g/\sqrt{2}$, so that we could have done the whole computation ignoring the distinction between e and g if we had also forgotten about counting how many fermions the external field couples to.

Now we must compute the energy of the vacuum due to the magnetic interaction in (13.85). In quantum mechanics we learn that the zero-point energy of a harmonic oscillator is $\frac{1}{2}\hbar\omega$. In quantum field theory, each energy eigentstate of the field acts like a harmonic oscillator whose zero-point energy is simply $\omega = \sqrt{p_z^2 + m^2 + (2n+1)eB}$, and the excited states have energy $(N + \frac{1}{2})\hbar\omega$, where the N here refers to the number of particles having energy ω, not to be confused with the Landau level of a given eigenstate. The zero-point energy density is given by the sum over all possible states,

$$U_{\text{dia}} = \frac{1}{2} \sum_n g_n \int \frac{dp_z}{2\pi} \sqrt{p_z^2 + m^2 + (2n+1)eB}, \qquad (13.86)$$

where g_n is the degeneracy of Landau levels per unit area of the x-y plane. We can easily deduce the value of g_n by taking the limit $B \to 0$, since then U_{dia} must reduce to the usual expression in the absence of any magnetic field,

$$U_{\text{dia}} \to \frac{1}{2} \int \frac{d^3 p}{(2\pi)^3} \sqrt{p_z^2 + m^2}. \qquad (13.87)$$

If we identify $(2n+1)eB = p_x^2 + p_y^2 \equiv \rho^2$, and $\int dp_x dp_y = 2\pi \int d\rho\rho$, then $2\rho\Delta\rho = 2neB$, and $\sum_n g_n = 2\pi \sum \rho\Delta\rho g_n/(2\pi eB) \to \int dp_x dp_y g_n/(2\pi eB)$. Comparison shows that therefore

$$g_n = \frac{eB}{2\pi} \qquad (13.88)$$

and

$$U_{\text{dia}}(B) = \frac{1}{2} \sum_n \frac{eB}{2\pi} \int \frac{dp_z}{2\pi} \sqrt{p_z^2 + m^2 + (2n+1)eB}. \qquad (13.89)$$

The sum (13.89) is obviously quite divergent in the UV, but we don't care about the most divergent part; we only care about the difference $U(B) - U(0)$ since the true vacuum contribution $U(0)$ is unobservable. To evaluate this difference, we can use the following identity:

$$
\int_{-\epsilon/2}^{(N+\frac{1}{2})\epsilon} dx\, F(x) = \sum_{n=0}^{N} \int_{(n-\frac{1}{2})\epsilon}^{(n+\frac{1}{2})\epsilon} dx\, F(x)
$$

$$
= \sum_{n=0}^{N} \int_{(n-\frac{1}{2})\epsilon}^{(n+\frac{1}{2})\epsilon} dx\, \left[F(n\epsilon) + (x-n\epsilon)F'(n\epsilon) + \frac{1}{2}(x-n\epsilon)^2 F''(n\epsilon) + \dots \right]
$$

$$
= \sum_{n=0}^{N} \left[\epsilon F(n\epsilon) + \frac{1}{24}\epsilon^3 F''(n\epsilon) + \dots \right] ;
\tag{13.90}
$$

therefore

$$
\sum_{n=0}^{N} \epsilon F(n\epsilon) = \int_{-\epsilon/2}^{(N+\frac{1}{2})\epsilon} dx\, F(x) - \frac{1}{24}\sum_{n=0}^{N} \epsilon^3 F''(n\epsilon) + \dots
$$

$$
= \int_{-\epsilon/2}^{(N+\frac{1}{2})\epsilon} dx\, F(x) - \frac{1}{24}\epsilon^2 \int_{-\epsilon/2}^{(N+\frac{1}{2})\epsilon} dx\, F''(x) + O(\epsilon^4)
$$

$$
= \int_{-\epsilon/2}^{(N+\frac{1}{2})\epsilon} dx\, F(x) - \frac{1}{24}\epsilon^2 F'(x)\Big|_{-\epsilon/2}^{(N+\frac{1}{2})\epsilon} + O(\epsilon^4).
\tag{13.91}
$$

We can apply this to $U(B) - U(0)$ using $\epsilon = 2eB$, $F(x) = (16\pi^2)^{-1}\int dp_z \sqrt{p_z^2 + x}$, to obtain

$$
U(B) - U(0) = \frac{1}{16\pi^2}\sum_n \int dp_z \left(-\frac{1}{24}\right)\frac{(2eB)^2}{2}[p_z^2 + m^2 + (2n+1)eB]^{-1/2}\Big|_{n=0}^{n=\infty}
$$

$$
= \frac{1}{24\cdot 8\pi^2}(eB)^2 \int_{-\Lambda}^{\Lambda} \frac{dp_z}{[p_z^2 + m^2 + eB]^{1/2}}
$$

$$
= \frac{1}{96\pi^2}(eB)^2 \ln(\Lambda/m),
\tag{13.92}
$$

which gives the diamagnetic contribution to the susceptibility

$$
\chi_{\text{diag}} = -\frac{1}{48\pi^2}e^2 \ln(\Lambda/m)
\tag{13.93}
$$

from each spin state. However, there is an implicit assumption in this calculation that we are dealing with bosons. Notice that

$$
U(0) = \frac{1}{2}\int \frac{d^3p}{(2\pi)^3}\sqrt{\vec{p}^2 + m^2}
\tag{13.94}
$$

is simply the contribution to the vacuum energy density which we know is positive for bosons and negative for fermions. Somehow our computation above misses the minus sign for fermions. Let's first see how (13.94) relates to our previous computations of the one-loop contribution to the vacuum energy density, which is

$$V = -\frac{i}{2} \int \frac{d^4p}{(2\pi)^4} \ln(-p^2 + m^2 - i\epsilon) .$$ (13.95)

It is easiest to compare dU/dm^2 and dV/dm^2. Note that

$$\frac{dU(0)}{dm^2} = \frac{1}{4} \int \frac{d^3p}{(2\pi)^3} (\vec{p}^2 + m^2)^{-1/2} ,$$ (13.96)

and letting $\omega = \sqrt{\vec{p}^2 + m^2}$, we can integrate over p_0 by completing the contour in the upper half plane to get

$$\begin{aligned}
\frac{dV}{dm^2} &= \frac{i}{2} \int \frac{d^3p}{(2\pi)^3} \int \frac{dp_0}{2\pi} \frac{1}{p^2 - m^2 + i\epsilon} \\
&= \frac{i}{2} \int \frac{d^3p}{(2\pi)^4} \frac{1}{2\omega} \int \frac{dp_0}{2\pi} \left(\frac{1}{p_0 - \omega + i\epsilon} - \frac{1}{p_0 + \omega - i\epsilon} \right) \\
&= \frac{i}{2} \int \frac{d^3p}{(2\pi)^4} \frac{1}{2\omega} (-2\pi i) .
\end{aligned}$$ (13.97)

This confirms (13.94), except that since we know V comes with a minus sign for fermions, it must also be the case for U.

Next we must consider the paramagnetic contribution to the energy, due to the intrinsic magnetic moment of the particles:

$$E^2 = \vec{p}^2 + m^2 + (2n+1)eB - ge\vec{B} \cdot \vec{S} \equiv E^2_{\text{Landau}} - ge\vec{B} \cdot \vec{S}$$ (13.98)

If we expand E in powers of the magnetic moment energy, we get

$$E = E_{\text{Lan.}} \left[1 - \frac{1}{2} \frac{geBS_z}{E^2_{\text{Lan.}}} - \frac{1}{8} \left(\frac{geBS_z}{E^2_{\text{Lan.}}} \right)^2 \right] ,$$ (13.99)

and we must sum the new contributions to $\frac{1}{2}\omega$ over all the Landau levels and p_z's. Picking out the term of order B^2 (note that the linear term will give no contribution when we sum over the spin polarizations) we get the paramagnetic contribution to the susceptibility,

$$U_{\text{para}} = -\frac{1}{2} \chi_{\text{para}} B^2 = -\frac{1}{16} (geS_z B)^2 \int \frac{d^3p}{(2\pi)^3 (p^2 + m^2)^{3/2}} ,$$ (13.100)

giving

$$\begin{aligned}
\chi_{\text{para}} B^2 &= \frac{1}{8} (geS_z B)^2 \int \frac{d^3p}{(2\pi)^3 (p^2 + m^2)^{3/2}} \times (\text{sign}) \\
&\cong \frac{4\pi}{8 \cdot 8\pi^3} (geS_z B)^2 \ln(\Lambda/m) \begin{pmatrix} +1, & \text{boson} \\ -1, & \text{fermion} \end{pmatrix} ,
\end{aligned}$$ (13.101)

where we used the same reasoning for the sign. The Landé g factor is always 2 for an elementary particle, at lowest order in perturbation theory—this is true for the gluons as well as the fermions. Putting the two contributions (paramagnetic and diamagnetic) together we obtain

$$\chi_{\text{total}} = \frac{1}{16\pi^2}\left[4S_z^2 - \frac{1}{3}\right]e^2\ln(\Lambda/m)\begin{pmatrix}+1, & \text{boson}\\ -1, & \text{fermion}\end{pmatrix}. \tag{13.102}$$

Comparing to (13.81), we can read off the contributions to the coefficient of the beta function,

$$\delta b_0 = -\frac{1}{32\pi^2}\left[4S_z^2 - \frac{1}{3}\right]e^2\ln(\Lambda/m)\begin{pmatrix}+1, & \text{boson}\\ -1, & \text{fermion}\end{pmatrix} \times (\text{no. of polarizations}). \tag{13.103}$$

First, let's check this result against QED. In that case, the photon carries no charge, so we get only the contribution from the fermions, which have $S_z = 1/2$. The way we have normalized the vacuum energy, each Dirac fermion has 4 polarization states (antiparticles must be counted separately), so this gives a beta function coefficient of $b_0 = +1/(12\pi^2)$, in agreement with (12.65).

Now let's look at the contribution from the vector gauge bosons in the nonAbelian theory. They have $S_z = 1$, and two polarization states. Therefore each virtual gluon which can couple to the external color B field contributes $-(11/3)/(16\pi^2)$ to b_0. Comparing to (13.69), we see that the form seems to be correct—we understand the factor of 11/3, and $C_2(A)$ is counting the virtual gluons. We can check that $C_2(A) = N$ is the correct factor in the context of our present computation by looking at the nonvanishing structure constants f^{abc} given in (13.9) for the case of QCD. Above we argued that we should compute the trace of the square of each generator which couples to the external field. For example, if we take the external field to be in the $a = 1$ color direction, we see that there are three generators which have this index, and

$$\sum_{bc} f^{1bc}f^{1bc} = 2(1 + 1/4 + 1/4) = 3, \tag{13.104}$$

which agrees with our expectation. We could of course have gotten the same answer using any value of a, as the identity (13.12) guarantees.

Thus we see that asymptotic freedom can be understood as a consequence of the fact that the gauge bosons themselves have magnetic interactions with an external B field, and being bosons they contribute with the opposite sign to the vacuum energy, hence the magnetic susceptibility. The virtual gauge bosons thus do cause antiscreening of a color charge, opposite to the screening behavior in QED.

Chapter 14
Nonperturbative Aspects of SU(N) Gauge Theory

To conclude this course, I would like to treat one important subject which goes beyond the realm of perturbation theory, namely the role of instantons in gauge theories. We will see that the vacuum state of QCD is much more complicated than that of QED, which is simply described by $A^\mu = 0$ or its gauge-equivalent copies. The reason for this is topology.

To motivate this, let's first consider a simpler model: QED in $1 + 1$ dimensions, where the spatial dimension is compactified on a circle. Let's parametrize it by $\theta \in [0, 2\pi]$. A possible gauge transformation is one that winds around the circle n times:

$$\Omega(\theta) = e^{in\theta}. \tag{14.1}$$

If we started from the vacuum state $A_\mu = 0$ and performed such a gauge transformation on the fermions, $\psi \to \Omega\psi$, then we would obtain a new value for the gauge field,

$$A_\mu \to -\frac{i}{e}\Omega\partial_\mu\Omega^{-1}; \quad A_0 = 0, \quad A_\theta = -\frac{n}{e}. \tag{14.2}$$

The interesting thing about this is that n has to be an integer; otherwise the gauge transformation is not continuous around the circle. Therefore it is not possible to transform it back to $A_\mu = 0$ using a family of gauge transformations which are continuously connected to the identity. This means that we would have to pass through some field configuration with nonvanishing field strength if we were to try to go between the different vacuum states.

The gauge transformations (14.1) with $n \neq 0$ are called *large* gauge transformations, and the vacuum states (14.2) are called n-vacua. Even though physics looks the same in any of these vacuum states, it is not obvious that we can choose just one of them. The "true" vacuum state could be a superposition of the different n-

---

The original version of this chapter was revised: The errors in this chapter have been corrected. The correction to this chapter can be found at https://doi.org/10.1007/978-3-030-56168-0_16

J. M. Cline, *Advanced Concepts in Quantum Field Theory*,
SpringerBriefs in Physics, https://doi.org/10.1007/978-3-030-56168-0_14

Fig. 14.1 A double well
potential

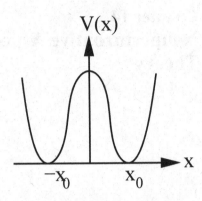

vacua. An analogous thing happens when we consider the quantum mechanics of
a double well potential, as in Fig. 14.1. Classically, there are two vacuum states,
at x_0 and $-x_0$. But quantum mechanically, the true vacuum state is the symmet-
ric superposition $|+\rangle = \frac{1}{\sqrt{2}}(|+x_0\rangle + |-x_0\rangle)$, which has a lower energy than the
antisymmetric one, $|-\rangle = \frac{1}{\sqrt{2}}(|+x_0\rangle - |-x_0\rangle)$. In quantum mechanics we learn
that this is related to the possibility of *tunneling* between the two states. The states
$|\pm x_0\rangle$ are not eigenstates of the Hamiltonian because the amplitude for tunnel-
ing is nonzero: $\langle +x_0| -x_0\rangle \neq 0$. Furthermore, the energy difference between the
symmetric and antisymmetric combinations is related to the tunneling amplitude:

$$
\begin{aligned}
E_+ - E_- &= \langle +|H|+\rangle - \langle -|H|-\rangle \\
&= \frac{1}{2}\left(\langle +x_0|H|-x_0\rangle + \langle -x_0|H|+x_0\rangle\right) \\
&\cong \langle \pm x_0|H|\pm x_0\rangle \langle -x_0|+x_0\rangle .
\end{aligned}
\tag{14.3}
$$

In field theory we expect a similar phenomenon whenever there are degenerate
vacuum states, as long as the probability to tunnel between them is nonvanishing. In the
above example, we found degenerate vacuum states as a consequence of topology: the
gauge transformations provided topologically nontrivial maps from the group manifold to physical space. In this example, the group manifold (for U(1)) is a circle, since
any two gauge transformations $e^{i\Lambda}$ and $e^{i\Lambda'}$ are equivalent if Λ and Λ' differ by a multiple of 2π. We mapped the group manifold onto another circle comprising the physical
space. In mathematical terms, this mapping has nontrivial *homotopy*. We could not
have done the same thing in $(2+1)$-D QED where the physical space is a sphere. In
this case, a gauge transformation which wraps around the equator of the sphere can be
smoothly deformed to a trivial gauge transformation at a point. There is no topological
barrier between large and small gauge transformations in this case (Fig. 14.2).

The preceding example makes it clear that if we want to find the same phenomenon in $3+1$ dimensions, we need a gauge group whose group manifold has
at least three dimensions. A simple example is SU(2): it has three generators, and
its group manifold is almost the same as that of SO(3), a three-sphere. To find the
large gauge transformations, let us imagine that 3-D space is compact (it could have

Fig. 14.2 A homotopically
trivial mapping

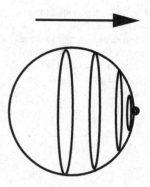

an arbitrarily large volume), and has the topology of a 3-sphere. We can therefore
assign coordinates ϕ_1, ϕ_2, ϕ_3 to physical space. Denote these collectively by $\vec{\phi}$. Then
the large gauge SU(2) transformations can be written as

$$U_n = \exp(in\vec{\phi} \cdot \vec{\tau}),\tag{14.4}$$

where $\tau_i = \frac{1}{2}\sigma_i$ are the generators of SU(2). We can find the nonAbelian gauge fields
generated from the trivial vacuum by these gauge transformations,

$$A_{n,\mu} = \frac{i}{g}U_n\partial_\mu U_n^\dagger.\tag{14.5}$$

Since they are pure gauge, they are guaranteed to have vanishing field strengths. As in
the Abelian example, the large gauge transformations (14.4) cannot be continuously
deformed into the trivial one $U = 1$ because of the topological obstruction: the
mapping U_n winds the group manifold around the physical space n times, and n
must be an integer by continuity of the fields.

We therefore expect that the true vacuum state of nonAbelian theories will be
some superposition of the n-vacua. In fact, it will turn out that the correct choice has
the form $|\theta\rangle = \sum_n e^{in\theta}|n\rangle$ where θ is an arbitrary number. But to understand this, we
have to show that the θ-vacua are the real eigenstates of the Hamiltonian, and this
requires knowing what is the amplitude for tunneling between any two n-vacuua.
How do we compute tunneling probabilities in quantum field theory? Instantons
provide the answer, at least in cases where there is a large barrier to tunneling. A
classic reference on this subject is [15].

14.1 Instantons

To find a transition amplitude in field theory, we can use the Feynman path integral.
We want to integrate over all field configurations which start in one vacuum state,
say $|n\rangle$, and end in the other, say $|m\rangle$. If the two states were separated by a saddle

Fig. 14.3 A potential for a
field with several
components, ϕ^i, and vacua at
the values $\vec{\phi}_m$ and $\vec{\phi}_n$. Dotted
line: path of smallest action
connecting the two vacua n
and m

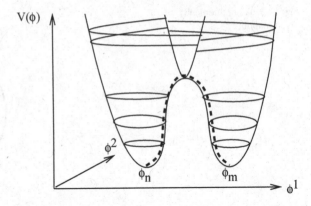

Fig. 14.3 A potential for a field with several components, ϕ^i, and vacua at the values $\vec{\phi}_m$ and $\vec{\phi}_n$. Dotted line: path of smallest action connecting the two vacua n and m

point as shown in Fig. 14.3, then there will be one path between them (the dotted line which goes over the saddle) which minimizes the action. This path will dominate the path integral in a stationary phase approximation:

$$\int \mathcal{D}\phi\, e^{iS} \sim e^{iS_{\min}} \tag{14.6}$$

This approximation will be justified if $S_{\min}$ is large, since then the action will have a large variation for nearby paths, and e^{iS} will oscillate rapidly, giving cancelling contributions to the path integral.

We can find the action of this path by solving the equation of motion:

$$-\ddot{\phi} + \nabla^2 \phi - V'(\phi) = 0, \tag{14.7}$$

subject to the boundary conditions $\phi(t \to -\infty) = \phi_n$, $\phi(t \to +\infty) = \phi_m$.

However, this method does not give a good approximation to the tunneling amplitude. The reason is that we should really integrate over all the fluctuations which are close to the stationary path in order to get a good estimate, and simply evaluating the integrand at the stationary point does not take this into account. We should see that the tunneling probability is exponentially small if the barrier is large.

There is a trick which enables us to do a very similar calculation to the one above, which however does give a good estimate of the tunneling amplitude. It is simply to go to Euclidean space before doing the path integral. When we do this, the path integral becomes

$$\int \mathcal{D}\phi\, e^{-S_E} \sim e^{-S_{\min}}. \tag{14.8}$$

We are now using the method of steepest descent rather than the stationary phase approximation to evaluate the path integral. Notice that this does immediately give what we expect: an exponentially suppressed amplitude in the case of a large barrier.

Fig. 14.4 The same potential as in Fig. 14.3, but now in Euclidean space

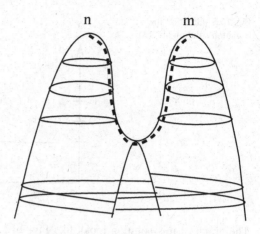

In Euclidean space, the action is given by

$$S_E = \int d^4x \left(\frac{1}{2} \left[(\dot{\phi})^2 + (\nabla \phi)^2 \right] + V(\phi) \right). \tag{14.9}$$

This is similar to the Minkowskian action, but with the potential reversed in sign. We can imagine we are still solving the Minkowskian field equations, but with $V \to -V$ (if we also lump the gradient term into the potential), as in Fig. 14.4.

The equation of motion becomes

$$\ddot{\phi} + \nabla^2 \phi - V'(\phi) = 0. \tag{14.10}$$

If we consider for a moment the case of quantum mechanics for a particle in a potential, the term $\nabla^2 \phi$ is absent since ϕ itself represents the position of the particle. The solution we seek is one that starts at n when $t \to -\infty$, and rolls to m as $t \to +\infty$. We can use energy conservation to write the first integral of the equation of motion,

$$E = 0 = \frac{1}{2}\dot{\phi}^2 - V(\phi) \tag{14.11}$$

where we chose the vacuum states to have zero energy for convenience. The solutions are given implicitly by

$$\int^t dt = \pm \int^\phi \frac{d\phi}{\sqrt{V(\phi)}} \tag{14.12}$$

For example, if $V(\phi) = \lambda(\phi^2 - v^2)^2$, the integral can be done, with the result

$$\phi(t) = v \tanh(\sqrt{\lambda} v t). \tag{14.13}$$

Fig. 14.5 Graph of the
instanton solution (14.13)

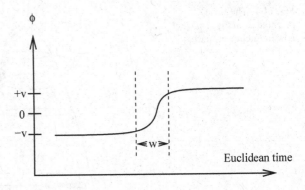

The graph of this solution looks like Fig. 14.5, where the width of the transition region is $w = (\sqrt{\lambda}v)^{-1}$. This is the amount of (Euclidean) time it takes for the field to do most of its transition between the two vacua, since for times much earlier or later, it is essentially just sitting at one of the two vacuua. Since all the action occurs at essentially one instant in time, 't Hooft called these kinds of solutions *instantons*. The journal Physical Review at first refused to accept such an imaginative name, so some of the first papers on this subject called them "pseudoparticles," but the instanton name quickly gained popularity.

An important feature of instantons is that they are *nonperturbative*. When (14.13) is substituted back into the Euclidean action, one finds that

$$S_{E,\min} \sim \frac{v^4}{\lambda}, \tag{14.14}$$

and therefore the tunneling amplitude goes like

$$\langle n|m \rangle \sim e^{-cv^4/\lambda}. \tag{14.15}$$

This has no expansion in perturbation theory—we would never see any sign of it even if we worked to infinite order in the perturbation expansion. Therefore instantons provide us with our first example of a truly nonperturbative effect in quantum field theory.

Now we are ready to apply these ideas to SU(2) gauge theory. We need a solution to the Euclidean Yang-Mills field equations

$$\partial^\mu F_{\mu\nu} + ig[A^\mu, F_{\mu\nu}] = 0 \tag{14.16}$$

which interpolates between two of the n-vacua (14.5) as t goes from $-\infty$ to $+\infty$. I will present this solution without proof, and leave it for you (last homework problem) to verify that it is a solution:

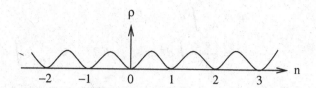

Fig. 14.6 Energy density of field configurations which interpolate between the n-vacua of Yang-Mills theory

$$A_\mu(x) = -\frac{i}{g}\frac{x^2}{x^2 + \lambda^2}U\partial_\mu U^\dagger, \qquad (14.17)$$

where $x^2 = x_0^2 + x_i^2$ and λ is an arbitrary constant length scale which characterizes the size of the instanton. The gauge transformation U is given by

$$U = \frac{1}{\sqrt{x^2}}(x_0 - i\vec{x}\cdot\vec{\sigma}). \qquad (14.18)$$

Clearly, as $x^\mu \to \infty$, $A_\mu(x)$ approaches a pure gauge field. But because of the prefactor $x^2/(x^2 + \lambda^2)$, it is not pure gauge in the interior, so there is a barrier. It turns out that the action of this solution is given by

$$S_I = \frac{8\pi^2}{g^2}, \qquad (14.19)$$

which like the scalar field example is nonperturbative.

The large gauge transformation U in (14.18) can be applied any number of times, $U_n = U^n$, to produce a map which wraps the group manifold n times around the 3-sphere at infinity. This means there is a series of vacuum states labeled by integers n, just as we had in the case of $(1+1)$-D QED on a circle. They are separated from each other by energy barriers in the field space, as shown in Fig. 14.6.

14.2   Winding Number

Since the instantons satisfy the Euclidean field equations, they are the field configurations which interpolate between n-vacuua with the least action. There are of course infinitely many other solutions which will have the same topology as a given instanton, but larger action. Is there a way in which we can compute whether a given field configuration has a nontrivial change in topology? I will not prove it, but one can show that the following expressions give the winding number of a gauge field in 2 and 4 Euclidean dimensions, for the U(1) and SU(N) gauge theories, respectively:

$$\nu = \frac{g}{4\pi} \int d^2x \, \epsilon_{\mu\nu} F_{\mu\nu};$$

$$\nu = \frac{g^2}{32\pi^2} \int d^4x \, \epsilon_{\mu\nu\alpha\beta} \, \mathrm{tr}\, F_{\mu\nu} F_{\alpha\beta} \equiv \frac{g^2}{16\pi^2} \int d^4x \, \mathrm{tr}\, F_{\mu\nu} \tilde{F}_{\mu\nu}. \qquad (14.20)$$

Both of these equations can be rewritten as integrals of a total divergence,

$$\nu = \frac{g}{2\pi} \int d^2x \, \partial_\mu W_\mu; \quad W_\mu = \epsilon_{\mu\nu} A_\nu;$$

$$\nu = \frac{g^2}{8\pi^2} \int d^4x \, \partial_\rho W_\rho; \quad W_\rho = \epsilon_{\rho\sigma\mu\nu} \, \mathrm{tr}\, \left(A_\sigma \partial_\mu A_\nu + \frac{2ig}{3} A_\sigma A_\mu A_\nu \right). \qquad (14.21)$$

If we compactify Euclidean space by imposing a maximum radius $\sqrt{x_\mu x_\mu} < R$ (we will take $R \to \infty$), then the divergence theorem can used to rewrite the above integrals as surface integrals, over the $(N-1)$-sphere which bounds the region $\sqrt{x_\mu x_\mu} < R$:

$$\nu = c_N \int d\Omega_{N-1}^\mu W_\mu, \qquad (14.22)$$

where $c_N = \frac{g}{2\pi}$ or $\frac{g^2}{8\pi^2}$. Now we can argue that any physically sensible configuration must have zero field strength at $r = R$ as $R \to \infty$, so that its action will be finite. This means that A_μ must approach a pure gauge configuration at the boundary,

$$A_\mu \to \frac{i}{g} U \partial_\mu U^\dagger. \qquad (14.23)$$

Now let's parametrize the bounding $(n-1)$-sphere by the angle ϕ in 2D and ϕ_1-ϕ_3 in 4D. The winding number (also called the Pontryagin index) becomes

$$\nu = -\frac{i}{2\pi} \int_0^{2\pi} d\phi \, \frac{dU}{d\phi} U^{-1}; \qquad (14.24)$$

$$\nu = \frac{1}{24\pi^2} \int d\phi_1 d\phi_2 d\phi_3 \, \epsilon_{ijk} \, \mathrm{tr}\, \left(\frac{dU}{d\phi_i} U^{-1} \frac{dU}{d\phi_j} U^{-1} \frac{dU}{d\phi_k} U^{-1} \right). \qquad (14.25)$$

It can be shown that these expressions are unchanged by deformations of $U(\phi_i)$ which do not change the winding properties of U; they are topological invariants. This is easy to see in the $U(1)$ case: suppose that $U = e^{if(\theta)}$ where f is any function which has the property that $f(2\pi) = f(0) + n2\pi$. Then (14.24) gives

$$\nu = \frac{1}{2\pi} \int_0^{2\pi} d\phi \, \frac{df}{d\phi} = \frac{f(2\pi) - f(0)}{2\pi} = n. \qquad (14.26)$$

The more complicated case of SU(N) in 4D works in a similar way. For example, consider the instanton solution (14.17). We can verify its winding number by considering the region near $\vec{x} = 0$ (the north pole of the 3-sphere). In this vicinity, choose the angles ϕ_i to be the space coordinates x_i, so

$$\partial_k U U^{-1} = i\sigma_k \qquad (14.27)$$

and

$$\text{tr}\left(\frac{dU}{d\phi_i} U^{-1} \frac{dU}{d\phi_j} U^{-1} \frac{dU}{d\phi_k} U^{-1}\right) = i^3 \text{tr}\sigma_i \sigma_j \sigma_k = 2\epsilon_{ijk}. \qquad (14.28)$$

Then $\epsilon_{ijk}\text{tr}(\cdots) = 12$ and we get

$$\nu = \frac{1}{2\pi^2}\int d\phi_1 d\phi_2 d\phi_3 = 1, \qquad (14.29)$$

since the volume of the 3-sphere (the group manifold of SU(2)) is $2\pi^2$. Here we pulled a little trick by assuming that the trace factor is a constant independent of the angles, even though we only showed it to be the case in the vicinity of the north pole. However, it can be shown that the expression for ν is invariant under constant tranformations of the group, $U \to UU_0$, so one can do the Faddeev-Popov procedure of inserting $1 = \int d\omega \delta(U - 1) \det|\partial U_\omega / \partial\omega|$. Since $\int dU \cdot 1 = 2\pi^2$ for this group, we deduce that $\int d\omega \det|\partial U_\omega / \partial\omega|_{U=1} = 2\pi^2$, which enables us to evaluate $\int dU \epsilon_{ijk}\text{tr}\left(\frac{dU}{d\phi_i} U^{-1}\frac{dU}{d\phi_j} U^{-1}\frac{dU}{d\phi_k} U^{-1}\right)$ just by knowing the value of the integrand in the vicinity of $U = 1$ (or any other element of the group, for that matter).

Let us comment on the relation of the instanton solution (14.17) and the picture we originally started with, that of tunneling in $(1 + 1)$-D QED compactified on a spatial circle. Logically, it is convenient to think about N-space being compactified on an N-sphere, and letting time run from $-T/2$ to $+T/2$, where we will take $T \to \infty$. Then we would look for gauge field configurations whose winding number on the N-spheres changes from n_- to n_+ as t goes from $-T/2$ to $+T/2$. The winding number of a $(N + 1)$-D instanton configuration that interpolates between the two vacuua can be shown to be

$$\nu = n_+ - n_-. \qquad (14.30)$$

However, it is not very convenient to look for instanton solutions on the space $S_N \times R$. It is much easier to find solutions that transform simply under SO(N+1) rotations of the full Euclidean space. (Remember, N is counting only the spatial dimensions.) Such solutions live on the $(N + 1)$-sphere rather than $S_N \times R$. However, in the end we are going to decompactify both spaces, so it should not matter which one we use, as long as the instantons have the same winding number in both pictures. In fact, one picture should be equivalent to the other under a gauge transformation as long as they both have the same winding number. The two situations are depicted in Fig. 14.7.

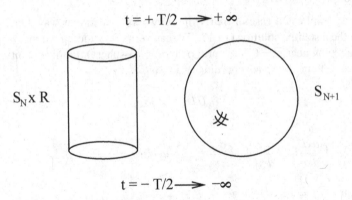

$t = + T/2 \longrightarrow + \infty$

$S_N \times R$ S_{N+1}

$t = - T/2 \longrightarrow -\infty$

Fig. 14.7 Two different ways of compactifying $N + 1$ dimensional Euclidean space

An important note for SU(N) gauge theories: although we have only displayed the instanton solution for SU(2), this is essentially all that is needed for SU(N) theories, since SU(2) is a subgroup of SU(N). For example, in SU(3), we can choose one of the 3 rows and columns to transform trivially, and embed the SU(2) instanton in the other two, for example

$$U_{\text{SU(3)}} = \begin{pmatrix} 1 & 0 \\ 0 & U_{\text{SU(2)}} \end{pmatrix}, \tag{14.31}$$

And any constant gauge transformation of this, $U \to U U_0$, is also a good solution. This means that the SU(3) instanton has an orientation in color space, over which we will have to integrate when we compute the tunneling rate.

14.3 Tunneling Between n-Vacuua

Previously we asserted that the amplitude for tunneling goes like e^{-S_I}, where the instanton action is given by $S_I = \frac{8\pi^2}{g^2}$. One way of writing this is in terms of the transition amplitude between between two neighboring n vacua with winding numbers n and $n + 1$. In Minkowski space the transition amplitude would be given by

$$\langle n + 1 | e^{iHT} | n \rangle, \tag{14.32}$$

since e^{iHt} is the evolution operator, but we are doing the calculation in Euclidean space, so we have

$$\langle n + 1 | e^{-HT} | n \rangle = \int \mathcal{D} A_\mu e^{-S} \sim C e^{-S_I}. \tag{14.33}$$

Here C is a prefactor which corrects the crude estimate e^{-S_I}. Coleman has explained very nicely how to compute C by expanding the action to quadratic order about the instanton and doing the path integral over these small fluctuations [15]. Not only does one have to sum over these small fluctations but also over the collective coordinates of the instanton. Namely, the instanton can be centered around any point in spacetime, and it can have different orientiations in color and in position space. Moreover, we saw that the instanton has an arbitrary size ρ which must also be integrated over. When all this is done, one gets a tunneling amplitude of

$$\langle n+1|e^{-HT}|n\rangle \sim VT \int \frac{d\rho}{\rho^5} e^{-8\pi^2/g^2(\mu)} f(\rho\mu). \qquad (14.34)$$

The factor VT is the volume of Euclidean spacetime, which tells us that the probability of tunneling should really be interpreted as a constant rate of tunneling per unit spatial volume (careful consideration of how the n-vacuum states should be normalized gives a tunneling rate that is proportional only to VT and not V^2T). The power of ρ could be guessed from dimensional analysis, as we will see below. Also $f(\rho\mu)$ is the function which arises from integrating over the small fluctuations. This is just some functional determinant, which is UV divergent, and requires us to renormalize, thus introducing the arbitrary renormalization scale μ. The μ-dependence of f is guaranteed to cancel that of $e^{-8\pi^2/g^2(\mu)}$. We can choose $\mu = 1/\rho$ so that f becomes an unimportant constant. Furthermore, let us remember that

$$g^2(\mu) \sim \frac{1}{2b_0 \ln(\Lambda/\mu)}, \qquad (14.35)$$

where Λ is the scale where the coupling becomes strong, on the order of 300 GeV in QCD. One then gets a tunneling rate of

$$\frac{\Gamma}{V} \sim \int \frac{d\rho}{\rho^5} e^{-8\pi^2/g^2(\mu)} \sim \Lambda^4 \int \frac{d\rho}{\rho} (\rho\Lambda)^{11N/3-4}, \qquad (14.36)$$

where we displayed the value of b_0 for pure SU(N) with no quarks. (The story actually becomes more complicated when we include quarks; see [15]). Notice that the powers of ρ are correct for giving a rate per unit time. Although the integral diverges for large instantons, it can be argued that our semiclassical approximation to the path integral is breaking down since their action is becoming too small, and the only reasonable place to cut off the integral is for instantons of size Λ^{-1}, since this is the only length scale in the theory. Thus one estimates finally that

$$\frac{\Gamma}{V} \sim \Lambda^4, \qquad (14.37)$$

which is no longer exponentially suppressed; this is a large value. So tunneling is a big effect, and we cannot pretend that the n-vacua are even good approximate vacuum states.

14.4 Theta Vacua

It turns out that the physically correct vacuum states are superpositions of the n-vacua of the form

$$|\theta\rangle = \frac{1}{\sqrt{2\pi}} \sum_n e^{-in\theta} |n\rangle. \tag{14.38}$$

Exactly the same thing happens in simple quantum mechanical systems with a periodic potential—these are the Bloch states. Unlike n-vacua, the theta vacuua are eigenstates of the large gauge transformations (14.4), U_p, since these simply increment the winding number of the n-vacua by p:

$$
\begin{aligned}
U_p|\theta\rangle &= \frac{1}{\sqrt{2\pi}} \sum_n e^{-in\theta} U_p|n\rangle \\
&= \frac{1}{\sqrt{2\pi}} \sum_n e^{-in\theta} |n+p\rangle \\
&= \frac{1}{\sqrt{2\pi}} \sum_n e^{-i(n-p)\theta} |n\rangle \\
&= e^{ip\theta}|\theta\rangle.
\end{aligned}
\tag{14.39}
$$

Furthermore, these states are orthonormal,

$$\langle\theta'|\theta\rangle = \frac{1}{2\pi} \sum_n e^{-in(\theta-\theta')} = \delta(\theta - \theta'), \tag{14.40}$$

and we can show that there is no tunneling between two different theta vacua:

$$
\begin{aligned}
\langle\theta'|e^{-HT}|\theta\rangle &= \frac{1}{2\pi} \sum_{n,m} e^{in\theta'-im\theta} \langle m|e^{-HT}|n\rangle \\
&\equiv \frac{1}{2\pi} \sum_{n,m} e^{in\theta'-im\theta} A(n-m) \\
&= \frac{1}{2\pi} \sum_n e^{in(\theta'-\theta)} \sum_{n-m} e^{i(n-m)\theta} A(n-m) \\
&= \delta(\theta - \theta') f(\theta).
\end{aligned}
\tag{14.41}
$$

Thus if we are in a particular theta vacuum, we will stay in it forever. This satisfies the criteria of a good vacuum state. One says that the theta vacua are different *superselection sectors*. Since there is no way to go from one to the other, it makes no sense to talk about superpositions of theta vacua.

Although all theta vacua seem to be acceptable vacuum states at this point, it is not true that the are all equivalent. In fact, we can show that they have different energies by computing the expectation value of e^{-HT},

$$\langle\theta|e^{-HT}|\theta\rangle = e^{-E(\theta)T}\langle\theta|\theta\rangle \cong (1 - E(\theta)T + \cdots)\langle\theta|\theta\rangle$$
$$\cong \delta(\theta - \theta)\sum_\nu e^{i\nu\theta}e^{-|\nu|S_i}$$
$$\cong VT\left(1 + 2\cos(\theta)e^{-S_i} + O(e^{-2S_i})\right). \tag{14.42}$$

We see that the vacuum energy density has the dependence

$$\frac{E(\theta)}{V} \sim \text{const.} - 2\cos\theta e^{-S_i}, \tag{14.43}$$

where the factor e^{-S_i} is just shorthand for the full expression which includes the integral over instanton sizes and thus gives the actual estimate

$$\frac{E(\theta)}{V} \sim \text{const.} - O(\Lambda^4)\cos\theta \tag{14.44}$$

Even though the vacua with $\theta = 0$ and π have the lowest energy, there is no way to get to these vacua if we happen to be in one with higher energy.

14.5 Physical Significance of θ

Strangely, it seems as though our discussion of the vacuum structure in SU(N) gauge theory has introduced a new coupling constant θ, on which physics seems to depends. How did we not see it when we originally formulated the theory?

In fact we can show that the θ parameter can be included at the level of the Lagrangian by adding the new term

$$\mathcal{L}_\theta = \frac{\theta g^2}{32\pi^2}\sum_a F_{\mu\nu}^a \tilde{F}_a^{\mu\nu}. \tag{14.45}$$

We originally discarded this term because it can be written as a total derivative (14.21), thus it has no effect on the equations of motion. In fact, the effects of such a term would vanish at any order in perturbation theory. However, we can now see that it does have a nonperturbative effect. Consider the expectation value of an arbitrary operator $\mathcal{O}$ in a theta vacuum. We have to sum over all the sectors of the gauge fields which have different winding number ν:

$$\langle \theta | \mathcal{O} | \theta \rangle = \sum_{\nu} e^{i\nu\theta} \int (\mathcal{D}A)_{\nu} e^{-S} \mathcal{O}$$

$$= \sum_{\nu} \int (\mathcal{D}A)_{\nu} e^{-S} \mathcal{O} \exp \left(i \frac{\theta g^2}{32\pi^2} \sum_{a} F_{\mu\nu}^a \tilde{F}_a^{\mu\nu} \right) \qquad (14.46)$$

where $(\mathcal{D}A)_{\nu}$ denotes the path integral restricted to gauge fields with winding number ν, and we used the relationship between winding number ν and $F\tilde{F}$ from (14.20). We see that the factor $e^{i\nu\theta}$ which weights the n vacuua is exactly reproduced by adding the term (14.45) to the QCD Lagrangian.

There is one big problem with this new term. It violates parity, and it can be shown that it would contribute enormously to the electric dipole moment of quarks and hence neutrons. There are stringent experimental limits which imply that $\theta \lesssim 10^{-10}$, a rather severe fine-tuning problem, known as the *strong CP problem*. Peccei and Quinn invented an elegant solution to this problem: by adding certain new fields, the theta parameter can be compensated by a dynamical field a, the axion. The potential for the axion becomes $E(\theta + a/f)$, where f is the axion scale (the scale of breaking of a new symmetry which is introduced), and the effective physical value of the theta parameter becomes $\bar{\theta} = \theta + a/f$. Since the axion naturally wants to go to the minimum of its potential, this gives a dynamical explanation for why $\bar{\theta} = 0$. Axions are still an attractive idea, appearing naturally in string theory, providing a candidate for dark matter, and they remain the subject of ongoing experimental searches.

Chapter 15
Exercises

1. (a) Prove that $s + t + u = m_1^2 + m_2^2 + m_3^2 + m_4^2$ for $2 \to 2$ scattering, where the four particles might all have different masses.
   (b) Compute the differential cross section $d\sigma/dt$ for $\psi\psi \to \psi\psi$ scattering for a Dirac fermion of mass m coupled to a scalar of mass μ via the Yukawa interaction $g\phi\bar{\psi}\psi$. Don't forget the u-channel diagram. For what scattering angle is $d\sigma/dt$ largest?

2. (a) Consider a theory with N Dirac fermions, labeled ψ_i with $i = 1, \ldots, N$, coupled to a scalar field via the interaction Lagrangian $g\phi \sum_i \bar{\psi}_i \psi_i$. Assuming the ϕ mass (μ) is much greater than the fermion masses, how does the ϕ decay width scale with N? How does the inclusive cross section for $\psi_1\bar{\psi}_1 \to$ all $\psi_j\bar{\psi}_j$ scale with N near the resonance, ignoring the width of the ϕ? How does the maximum value of the cross section scale with N, accounting for the width of the ϕ?
   (b) Use the result you found in (a), along with the result for a single species of fermions derived in the lectures, to estimate the maximum value of the cross section for $Z \to$ all $q\bar{q}$ (i.e., quarks and antiquarks). The number of decay channels for the Z boson is 3 (colors) for every flavor of quark which is kinematically allowed, plus 1 for each type of charged lepton and neutrino. Compare your answer to the actual value, which you can find in the section on "Plots of cross sections and related quantities" of the Particle Data Group Review of Particle Physics. (See http://pdg.lbl.gov/2004/reviews/contents_sports.html)
   (c) The comparison between the toy model prediction and the data is not very good in part (b). One observation that helps is that the Z boson is not a scalar particle, but rather a massive vector, with three polarization states. How will your prediction in (b) change if you treat each of these polarization as though they were separate scalar particles?

3. Study Eq. 10.47 of the section "Electroweak model and constraints on new physics" in the Particle Data Book. (a) What would be the effective value of g^2 for each kind of quark and lepton, if the Z boson was a scalar particle? Note that

The original version of this chapter was revised: The errors in this chapter have been corrected. The correction to this chapter can be found at https://doi.org/10.1007/978-3-030-56168-0_16

J. M. Cline, *Advanced Concepts in Quantum Field Theory*,
SpringerBriefs in Physics, https://doi.org/10.1007/978-3-030-56168-0_15

the partial widths in (10.47c) for each quark are summed over 3 colors, so you should divide these results by 3. Ignore QED and QCD corrections.

(b) Sum all the partial widths, assuming that charmed and strange quarks are like up and down quarks. Do they add up to the total width (10.48)? Note: if you are unsure about masses and charges of quarks and leptons, see the summary tables at http://pdg.lbl.gov/2004/tables/contents_tables.html.

4. Consider a model of two scalar fields with potential $M^2\sigma^2/2 + m^2\phi^2/2 + \mu\sigma\phi^2/2$. Assume that the decay $\sigma \to 2\phi$ is allowed.

    (a) Compute the decay width of the σ.

    (b) Compute the imaginary part of the σ self-energy. Show that it is related to the decay width in (a) in the manner claimed in lecture, near the σ mass shell. (This form is also mentioned at the end of the kinematics review in the Review of Particle Properties).

    (c) Evaluate the most divergent terms, and determine the counterterms in the tree level Lagrangian needed to cancel these divergences.

5. Evaluate the self-energy correction to the fermion $\Sigma_\psi(p)$ which was started in lecture. First do all integrals except the Feynman parameter integral. For the last step, compute only the divergent part.

6. Show that the series of one-particle reducible diagrams which contribute to the two-point function in the theory of a single scalar field with potential $m\phi^2/2 + \mu\phi^3/3!$ is a geometric series. (You should see the pattern for the statistical factor after computing it for the first few diagrams.) Evaluate the most divergent contribution and sum the series. Why is it important that the series is geometric?

7. Consider a scalar field whose potential energy is $V(\phi) = A\phi^6$.

    (a) What is the tree-level mass of the corresponding particle? What is the value of the cross section for $2\phi \to 2\phi$ at tree level?

    (b) What is the dimension of the coupling constant A?

    (c) Derive the invariant amplitude $\mathcal{M}$ for the process $2\phi \to 4\phi$ at lowest order in A.

    (d) Compute the lowest order quantum correction to the mass of the particle.

    (e) Compute the lowest order quantum correction to the differential cross section $d\sigma/dt$ for $2\phi \to 2\phi$.

    (f) Draw the connected Feynman diagrams contributing to the two-loop correction to the 6-point function $G_6(p_1, \ldots, p_6)$. Identify those which are 1PI or 1PR.

8. Consider the one-loop vertex correction $\Gamma_{\phi\bar\psi\psi}$ in the Yukawa theory ($V = -g\phi\bar\psi\psi$), in the limit where the fermion mass vanishes, and the external momentum of the boson vanishes, but the fermion momentum p and the boson mass μ are nonzero. Use a momentum-space cutoff Λ and drop all terms which vanish as $\Lambda \to \infty$, but make no other approximations.

    (a) $i\Gamma_{\phi\bar\psi\psi}$ has a real and an imaginary part. Compute the real part (which corrects the tree-level coupling). Note that $\int dx \ln(Ax^2 + Bx + C)$ can be computed by completing the square in the logarithm. For definiteness, assume that $p^2 > \mu^2$.

(b) Verify that the effective coupling depends on p^2 at large p^2 in the way claimed in the lecture about the running of the coupling constant, $g(\mu_r)$.

(c) For what values of p^2 does $i\Gamma_{\phi\bar{\psi}\psi}$ develop an imaginary part? Show that these are the same values as those for which the virtual particles in the loop can go on shell. Notice the similarity to the other situation where you have seen the loop get an imaginary part. The imaginary part is related to unitarity of the S-matrix through the optical theorem.

9. Consider the scattering process $2\phi \to 2\phi$ in the theory with couplings $\lambda\phi^4/4! + g\phi\bar{\psi}\psi$, where the external particles all have energy E which is much greater than any of the masses.

(a) Explain why, at such large energies, any loop corrections (including finite part) containing logarithmic divergences should go like $\ln(\Lambda/E)$, rather than $\ln(\Lambda/\mu)$, for example.

(b) Use the observation of (a) to deduce the energy dependence at high E of the one-loop amplitude for $2\phi \to 2\phi$ scattering, by computing only the most divergent contributions. Does the effective coupling $\lambda(E)$ grow or decrease as a function of E?

10. Consider the nonrenormalizable theory with potential $M^2\phi^2/2 + \lambda M^{-2}\phi^6/6!$. Suppose that M is some fixed mass scale which is *not* considered to be a function of the cutoff, but instead is just a generic value which is parametrically smaller than the cutoff. Assume that $\lambda \lesssim 1$. For this problem think of the cutoff as being some actual physical scale, large but not infinite.

(a) A given 1PI Feynman diagram will have $\mathcal{V}$ vertices and $\mathcal{E}$ external legs. How many internal lines I must there be?

(b) Prove that the number of loops is $\mathcal{L} = I - V + 1$.

(c) Estimate the contribution of this diagram to the proper vertex with $\mathcal{E}$ legs, $\Gamma_{\mathcal{E}}^{(V)}$, as a function of $\mathcal{V}$ and $\mathcal{E}$. Take the external legs to be at zero momentum. Assume combinatoric factors will approximately cancel out the factors of $1/6!$, and don't worry about minus signs, but keep factors of 2π. Estimate the loop integrals by power counting: let $\prod_{i=1}^{\mathcal{L}} d^4 l_i \to dl\, l^{4\mathcal{L}-1}$ and $\prod_{j=1}^{I}(l_j^2 + M^2) \to (l^2 + M^2)^I$. Will the integral be dominated by infrared (low l) or ultraviolet (high l) contributions, for a given value of $\mathcal{E}$ and $\mathcal{V}$?

(d) By summing the dominant contributions for a given $\mathcal{E}$, estimate the size of full proper vertex, $\Gamma_{\mathcal{E}} = \sum_{\mathcal{V}} \Gamma_{\mathcal{E}}^{(\mathcal{V})}$.

(e) In terms of the proper vertices, write down the form of an effective potential $V_{\text{eff}}(\phi)$ which has the property that tree level diagrams computed from that potential reproduce the full multiloop result.

(f) Operators of dimension $d < 4$ are known as *relevant*, with $d = 4$ as *marginal*, and with $d > 4$ as *irrelevant*. From the above results, explain why this is an appropriate terminology, when the cutoff is large. This represents the modern understanding of what is special about renormalizable theories (which contain only relevant and marginal operators): the nonrenormalizable operators have small coefficients.

11. This problem is supposed to solidify your intuitive understanding of renormalization, and give you practice with the Wilsonian effective action. Consider the
theory with potential $V = A\phi + \frac{1}{2}m^2\phi^2 + \frac{1}{6}\mu\phi^3$ in 6 spacetime dimensions.
(a) What are the dimensions of the couplings and the field?
(b) Show by power counting of the loop diagrams which are generated that the
theory is renormalizable.
(c) The tadpole term $A\phi$ can be removed by redefining the field $\phi = \phi_0 + \delta\phi$,
provided that the parameters satisfy a certain relation. Find $\delta\phi$ and the relation.
Show how the physical mass m_0^2 and cubic coupling μ_0 depend on the original parameters after going to the new field ϕ_0. Show that the value of $\delta\phi$ and
the shifted couplings could equivalently be found by imposing that $V'(\phi_0) = 0$,
$m_0^2 = V''(\phi_0)$ and $\mu_0 = V'''(\phi_0)$.
(d) We need to do loop integrals in 6 dimensions in the sections below. In d dimen-
sions, let $d^d x = d\Omega_{d-1}\, dx\, x^{d-1}$. Using the fact that $\int_{-\infty}^{\infty} dx\, e^{-x^2/2} = \sqrt{2\pi}$,
evaluate the integral $\int d^d x\, e^{-x^2/2}$ in both Cartesian and spherical coordinates to
determine $\int d\Omega_{d-1}$.
(e) Let $A = 0$ at the cutoff Λ. Write down integral expressions for the terms in
$i\delta S_\Lambda \equiv i S_{\Lambda'} - i S\Lambda$ which will contribute to renormalization of the original couplings in S_Λ when computing Wilson's effective action at a lower cutoff $\Lambda' < \Lambda$.
(Do not compute contributions to nonrenormalizable operators.) By using
the shorthand $\int_{\Lambda'}^{\Lambda} dp = \int d^d p/(2\pi)^d$ (integrated between Euclidean momenta
$\Lambda' < |p| < \Lambda$), $\delta_q = (2\pi)^d \delta^{(d)}(\sum_i q_i)$ and $\phi_q = \int d^d x\, e^{iqx}\phi(x)$, write these
contributions to $S_{\Lambda'}$ in the form of integrals which do not make any reference to
the dimensionality of spacetime. Use q_i for the external momenta and p for the
loop momentum in your expressions.
(f) Now specializing to $d = 6$, evaluate the most divergent parts of the integrals
(from Λ' to Λ) in part (e). For the self-energy diagram, Taylor expand in the
external momentum so that you can identify both the mass and the wave function renormalization. Why is it reasonable to neglect higher terms in the Taylor
expansion? Assemble your results into a complete expression for these leading
contributions to $i\delta S_\Lambda$, rewritten in position space rather than momentum space.
(g) Combine δS_Λ with S_Λ to find $S_{\Lambda'}$ and identify the corrections to the kinetic
term and the couplings. Canonically normalize the field at the new cutoff to
absorb the wave function renormalization found in (f). What are the new values
of $A(\Lambda')$, $m^2(\Lambda')$ and $\mu(\Lambda')$ in terms of the original parameters? Keep in mind
that you are doing a perturbative expansion.
(h) How does the cubic coupling run with Λ'? By setting $\Lambda' = E$ for an experiment at energy scale E, decide whether scattering processes become stronger or
weaker as you go to higher energies. Compute the beta function for this coupling,
$\beta(\mu) = \Lambda' \frac{d}{d\Lambda'} \mu(\Lambda')$ and take note of its sign. Would it have been possible to
deduce this result from just one of the Feynman diagrams you calculated?
(i) Find the physical $m^2(\Lambda')$ which takes into account the shift in ϕ needed to
remove the tadpole. Then evaluate it as $\Lambda' \to 0$, and take this to be the physical
mass squared, m_{phys}^2. Keep only the terms which dominate at small coupling

and large cutoff. Assuming $m^2_{\text{phys}} \ll \Lambda$, solve for the value of $m^2(\Lambda)$ needed to obtain the small m^2_{phys}, to first order in m^2_{phys}, and show that there is a fine-tuning (hierarchy) problem.

(j) Estimate the magnitude of the leading contribution to the energy density of the vacuum (*i.e.*, cosmological constant) which is generated at the scale Λ', assuming it was zero at the scale Λ.

In the following two problems you will solve the renormalization group equations for the theory with potential $\frac{1}{2}\mu^2\phi^2 + m\bar{\psi}\psi + \lambda\phi^4/4! + g\phi\bar{\psi}\psi$.

12. (a) Verify the result given in lecture for the anomalous dimensions of ϕ and ψ at one loop: $\gamma_\phi = g^2/8\pi^2$ and $\gamma_\psi = g^2/32\pi^2$.

(b) Show that this gives the beta function $\beta(g) = 5g^3/16\pi^2$.

(c) Use the anomalous dimensions and the vertex correction you computed in a previous assignment to find the contribution from the fermion loops to the beta function of the quartic coupling. (I was incorrect to say in lecture that this contribution β_λ has to be positive.)

13. (a) Using Mathematica or some other programming language, numerically integrate the beta functions to obtain the running couplings as a function of $t = \ln(\mu_r/\mu_0)$. (Don't forget to use the full beta function for λ.) Start with $\lambda = 0.5$, $g = 0.4$ at $t_0 = 0$. At what value of t does the quartic coupling vanish? How many orders of magnitude increase in the energy does this value of t represent? Plot the two couplings as a function of t.

(b) Keeping $\lambda(0) = 0.5$, by trial and error find the smallest value of $g(0)$ where a Landau pole occurs close to $t = 100$. Find its value to 3 significant figures. Then show that for larger values of $g(0)$, it is possible to keep both couplings in the range $[0, 1]$. For what range of values of $g(0)$ is this true? If you had to make an analytic estimate of the best value of $g(0)$ to take, for a fixed value of $\lambda(0)$, what would it be? Compare to your numerical result.

(b) Add the anomalous dimension for ϕ in your system of equations. Use this together with your determination of λ to plot the scale dependence of the amplitude for $\phi\phi \rightarrow \phi\phi$. On the plot show both the running coupling by itself, and the amplitude including the effect of the anomalous dimension. Does the anomalous dimension increase or decrease the amplitude? Do it first for the set of couplings in (a) which give the Landau pole, but graph only the region of t for which the couplings remain perturbative. Do it again for the second set of couplings, graphing over the whole range of t.

14. Evaluate $\frac{1}{2}\int d^4x\, d^4y\, J^\mu(x)D_{\mu\nu}(x-y)J^\nu(y)$ for the static charge distribution $J_0(x) = \rho(\vec{x})$, $J_i = 0$, and show that it gives the self-energy of the charge distribution due to the Coulomb potential, times an infinite integral over time (which converts the energy into the action). Remember to use the appropriate $i\epsilon$ prescription to define the poles of the propagator. Do the q_0 integral of the propagator first, as a contour integral, which must be done separately for the two cases $x_0 < y_0$ and $y_0 < x_0$. Next do the x_0 integral. Then do the angular integral of the propagator, and finally the radial momentum integral of the propagator.

15. In lecture it was claimed that the gauge fixing term does not spoil the gauge invariance of any other terms in the effective action. Thus the α-dependence which enters these terms must do so in a gauge-invariant way. To see how this works, compute the divergent and α-dependent parts of the electron self-energy, vacuum polarization, and vertex correction at one loop in QED (using a general Lorentz gauge), and show that they are consistent with the Ward identities. Verify that the beta function for the charge is independent of α.

16. Denote the two-fermion, n-photon terms in the QED effective action by

$$\Gamma_{\bar\psi A^n \psi} = \sum_{n=0}^{\infty} \int \prod_{i=0}^{n} dp_i \, \bar\psi(p_{n+1}) \Gamma_{\mu_1...\mu_n} \psi(p_0) A^{\mu_1}(p_1) \cdots A^{\mu_n}(p_n) \delta_{p_{n+1}-\sum p_i}$$

(15.1)

(where $dp = d^4p/(2\pi)^4$ and $\delta_{p_{n+1}-\sum p_i} = (2\pi)^4 \delta^{(4)}(p_{n+1} - \sum_{i=0}^{n} p_i)$). Find the Ward-Takahashi identities which relate the functions $\Gamma_{\mu_1...\mu_n}$. Draw the 1-loop diagrams corresponding to these identities (at one-loop level) to illustrate.

17. Derive the axial anomaly using the point-splitting regularization of the axial vector current. Use the following steps. (a) Expand the exponential of the line integral to order ϵ_μ and show that to this order,

$$\partial^\mu J_{\mu 5} = ieF^{\mu\nu}\epsilon_\nu \bar\psi(x+\epsilon/2)\gamma_\mu\gamma_5\psi(x-\epsilon/2)$$

(15.2)

(b) Show that the vacuum expectation value of the above quantity vanishes at zeroth order in perturbation theory. Then set up the calculation of the same quantity at first order in perturbation theory in e (one insertion of the electromagnetic interaction in a background gauge field).

(c) Let p and q be the momenta of the two fermion propagators, and let $l = p - q$ be the momentum of the gauge field which you have inserted. Further, let $r = p + q$. Show that your result is proportional to

$$\int d^4l \int d^4r \, \frac{e^{il\cdot x - i\epsilon\cdot r/2} l_\alpha r_\beta A_\nu(l)}{((r+l)^2/4 - m^2)((r-l)^2/4 - m^2)}$$

(15.3)

Since we are interested in the $\epsilon \to 0$ behavior, argue that we can set $l \to 0$ when doing the r integral. You should now see how to do the l integral.

(d) To do the r integral, note the following useful trick: rewrite $1/(r^2)^2$ as $\int_0^\infty ds\, s\, e^{-sr^2}$, and $r_\beta = 2i\frac{\partial}{\partial\epsilon^\beta}$.

(e) Average over the directions of ϵ_ν to evaluate $\epsilon_\mu\epsilon_\nu/\epsilon^2$ and obtain the final result.

18. (a) In $D = 2$ dimensions, we can define Dirac matrices

$$\gamma_0 = \sigma_x, \quad \gamma_1 = i\sigma_y, \quad \gamma_5 = -\sigma_z$$

(15.4)

Show that they satisfy the usual Dirac algebra for the anticommutators, and also that $\text{tr}\gamma_\mu\gamma_\nu\gamma_5 = 2\epsilon_{\mu\nu}$.

(b) Use the above results to do Fujikawa's derivation of the axial anomaly in 2D. To remind you, the steps are: (1) Show that the Jacobian in the path integral for an infinitesimal $U(1)_A$ transformation has the form $\det(1 + i\epsilon(x)\gamma_5)$ for either ψ or $\bar{\psi}$, where the determinant is in the space of all functions as well as 2D spinor indices. (2) Using the identity $\ln\det = \operatorname{tr}\ln$, rewrite this determinant in a more explicit form involving $\int d^4x \int d^4p$, and regulate the momentum space integral in a way which respects the $U(1)$ gauge symmetry, i.e., using a function $f((i\!\!\!D)^2/\Lambda^2)$, where D is the covariant derivative, Λ is a UV cutoff scale, and f is a function with the properties $f(0) = 1$, $f(x) \to 0$ as $x \to \infty$. (3) Use the result which is analogous to $D^2 = D^2 - (i/2)F_{\mu\nu}\gamma^\mu\gamma^\nu$ (is this still valid in 2D?) and expand f to first order in $F_{\mu\nu}\gamma^\mu\gamma^\nu$. (4) Show that the first order term is sufficient to get a nonvanishing result for the trace. Rewrite the momentum integral in the form $\int dp^2\, p^2 f'(p^2)$. Use integration by parts to evaluate the momentum integral, showing that the result for the anomaly is independent of the detailed form of the cutoff function f.

(c) Based on the 2D and 4D results, guess the form of the anomaly in $2n$ dimensions. Why is there no axial anomaly in $2n + 1$ dimensions?

19. In lecture it was shown that the amplitude for $\pi^0 \to \gamma\gamma$ due to the triangle anomaly is

$$A(\pi^0 \to \gamma\gamma) = \frac{N_c e^2}{4\pi^2 f_\pi}(Q_u^2 - Q_d^2)\,\epsilon_{\mu\nu\alpha\beta}\,p_1^\alpha\,p_2^\beta\,\epsilon_1^\mu\,\epsilon_2^\nu$$

From this compute the decay width. Compare to the experimental value $\Gamma = 8.0$ eV to determine the combination $N_c\,(Q_u^2 - Q_d^2)$. Using the known values of the quark charges, find the number of colors, rounded to the closest integer.

20. (a) Compute the divergent contributions to the gluon vacuum polarization in QCD due to diagrams involving ghosts and gluons, in Feynman gauge. Find the contribution to the Z_3 renormalization factor of QCD coming from these diagrams. Notice that this is the complete answer in a theory with just gauge bosons and no quarks.

Hint: think about the value of a diagram which has no external momentum flowing through the loop, in a massless theory using dimensional regularization, before computing group theory factors for such a diagram.

(b) To find the beta function, one must combine this result with that of other diagrams, since it is gauge-dependent. The simplest way to proceed for the pure gauge theory is to evaluate Z_1/Z_2 in a theory with quarks. The result in Feynman gauge is

$$Z_1/Z_2 = 1 - \frac{g^2}{16\pi^2\epsilon}N \tag{15.5}$$

where $N = 3$. (Only in axial gauge where the ghosts decouple is $Z_1/Z_2 = 1$, and then the complete result can be gotten from Z_3 as in QED.) Using this result and that you found in part (a), compute the beta function for the coupling in the pure gauge theory.

Correction to: Advanced Concepts in Quantum Field Theory

Correction to:
J. M. Cline, *Advanced Concepts in Quantum,*
Field Theory, **SpringerBriefs in Physics,**
https://doi.org/10.1007/978-3-030-56168-0

In the original version of the book, author-provided belated corrections have been incorporated as follows:

In Chap. 1, corrections have been incorporated in the third paragraph of Page 1 and in Figs. 1.4, 1.5 and 1.6.

In Chaps. 6 and 7, punctuation has been added to some of the equations.

In Chap. 9 corrections have been incorporated in Fig. 9.5.

In Chap. 10, corrections have been incorporated in the fifth paragraph of Page 69.

In Chap. 11, spacing issue in Equation (11.23) has been corrected.

In Chap. 12, spacing issue in Equations (12.7) and (12.9) and content in the fourth paragraph of Page 96 have been corrected.

In Chap. 13, content in the first paragraph of Page 114 has been corrected.

In Chap. 14, spacing issues in Equations (14.17) and (14.18) have been corrected.

In Chap. 15, the content in the last paragraph in Exercise 3 has been corrected.

Numbers have been added to all the references. The book and the chapters have been updated with the changes.

The updated version of the book can be found at 10.1007/978-3-030-56168-0

References

1. Bjorken, Drell, *Relativistic Quantum Fields* (McGraw-Hill)
2. Itzykson, Zuber, *Quantum Field Theory* (McGraw-Hill)
3. Churchill, Brown, Verhey, *Complex Variables and Applications*, 3rd edn. (McGraw-Hill)
4. https://pdg.lbl.gov/2020/reviews/rpp2020-rev-kinematics.pdf
5. P. Ramond, *Field Theory: A Modern Primer* (Frontiers in Physics)
6. D.J. Gross, *Applications of the renormalization group, in Methods in Field Theory*. Les Houches Lectures (1975)
7. J.P. Preskill, unpublished lecture notes on QFT and QCD. http://theory.caltech.edu/~preskill/notes.html
8. K.G. Wilson, J. Kogut, Phys. Rep. **12C**, 76 (1974)
9. L.F. Abbott, Nucl. Phys. B **185**, 189 (1981)
10. S.R. Coleman, E. Weinberg, Phys. Rev. D **7**, 1888 (1973)
11. E.S. Abers, B.W. Lee, Phys. Rep. **9**, 1 (1973)
12. K. Fujikawa, Phys. Rev. Lett. **42**, 1195 (1979)
13. J. Schwinger, Phys. Rev. **128**, 2425 (1962)
14. Peskin, Schroeder, *An Introduction to Quantum Field Theory* (CRC Press)
15. S. Coleman, "The Uses of Instantons," 1977 Erice lectures in "Aspects of Symmetry: Selected Erice Lectures", (Cambridge University Press)

© The Author(s), under exclusive license to Springer Nature Switzerland AG 2020
J. M. Cline, *Advanced Concepts in Quantum Field Theory*,
SpringerBriefs in Physics, https://doi.org/10.1007/978-3-030-56168-0

Printed in the United States
by Baker & Taylor Publisher Services